Underground Empires

Two Centuries of Exploration, Adventure
& Enterprise in New York's Cave Country

To Mike —

Dana Cudmore

Published by
Black Dome Press Corp.
PO Box 64, Catskill, NY 12414
blackdomepress.com
518.577.5238

First Edition Paperback 2021

Copyright © 1990 by Dana Cudmore
Copyright © 2005 by Dana Cudmore
Copyright © 2021 by Dana Cudmore

Portions of this book were originally published by Overlook Press in 1990 as *The Remarkable Howe Caverns Story* and by Media Services in 2005 as *Unearthing Howes Cave: A Community and a Quarry from 1842 On*.

Without limiting the rights under copyright above, no part of this publication may be reproduced, stored in or introduced into a retrieval system, or transmitted, in any form, by any means (electronic, mechanical, photocopying, recording, or otherwise), without the prior written permission of the publisher of this book.

ISBN: 978-1-883789-98-5

Library of Congress Control Number: 2021942087

Front cover: An 1889 S.R. Stoddard photo of a tour group in Howe Caverns' "Washington's Hall"; (inset) one of the early iterations of the third Cave House to welcome visitors to Howe's Cave. Built of cut stone in about 1872, the once abandoned hotel is now being revitalized as a museum of mining and geology.

Back cover (left to right): a photo from the early 1900s shows "Signature Rock," located in an undeveloped part of Howe Caverns beyond the Lake of Venus; a large, hand-painted bat says goodbye to visitors as they leave Secret Caverns; 1890s photo of workers with an air drill in the cement quarry in Howes Cave.

Design: Toelke Associates, www.toelkeassociates.com

Printed in the USA

10 9 8 7 6 5 4 3 2 1

Underground Empires

Two Centuries of Exploration, Adventure & Enterprise in New York's Cave Country

Dana Cudmore

BLACK·DOME
blackdomepress.com

Dedication

It's hard to believe that it was more than thirty years ago that I dedicated my first book on Howe's Cave to my three young daughters—Libby, Hilary, and Laura. They have since grown, of course, and married. Libby, another author in the family, is writing dedications to her own books. She and her husband, Ian, enjoy life not far away. Hilary and Bill have two girls—Lucy and Melody; Laura and Chris have boys—Max and Zack. This book is dedicated with love to all.

Much of this book was written during the COVID-19 pandemic, and family visits were sadly limited. In the years ahead I look forward to sharing some of the stories here with my grandkids (like I did with my own children)[1] and taking them to some of the key locations around the fascinating Schoharie County cave country.

I want to also recognize the many, many friends I made as a young man working summers at Howe Caverns and exploring some of Schoharie County's caves. As it was with hundreds of teenagers over the years, guiding at the cave was my first "real" job and I looked forward to it every year. For that reason this book is also dedicated to the current staff and tour guides at Howe Caverns and Secret Caverns, as well as to cavers everywhere.

---

1. To this day my daughters can still point out cave locations along the back road from Cobleskill to the church in Carlisle we attended.

Contents

Foreword	viii
Map	xii–xiii
Acknowledgments	xviii
Preface	xxi

Section I: The Remarkable Howe Caverns Story

Introduction: *A History with Parallels to the Growth of a Nation*   2

1. **Earliest Records**   5
   A Short History of the Schoharie Valley; Indian Legends and a Revolutionary War Hideout; the Story Behind Lester Howe's 1842 Discovery

2. **The Rival of Mammoth Cave**   21
   Led by Torchlight, Tourists Spend Eight Hours Underground; "Pip's Journal"

3. **Lester Howe Loses His Cave**   38
   "Lester Howe . . . is not a very Smart Man"; Corporate Maneuverings and Quarry Operations Close Howe's Cave

4. **They Bored a Hole in a Hill**   64
   The Tragic Story of Floyd Collins; a Bold Plan for Reopening Howe's Cave

5. **Work Begins Underground**   77
   The Engineering Story; Workers Trapped by Sudden Flooding; Roger Mallery, Workers' hero

6. **A Grand Reopening**   91
   Triumph and Tragedy

7. **The Opening of Secret Caverns and Knox Cave**   107
   Success Encourages Competition

8. **The Search for the Garden of Eden**   129
   The Legacy of Lester Howe

9. Modern History, 1929–1990   144
   Lester Howe's Descendants Visit the Modern Cave; a Brief Geology Lesson on Howe Caverns and Secret Caverns

10. Going Underground   153
    "A Romantic Trip" through Howe; a No-Frills Trip through Secret Caverns

Section II: Unearthing Howes Cave

The Community and the Quarry from 1842 On

1. A Famous Discovery   172
   More than Just a Cave

2. The Industry Grows   182
   The Mine Moves above Ground; the Cave House Burns

3. Boom Times Ahead   189
   A Two-Room Schoolhouse, Two Churches, and a Chevy Dealership

4. Tales of Tite Nippen   196
   Where'd That Name Come From?

5. Hard Times   201
   Competition and Environmental Challenges

6. Some Historic Firsts   205
   A Chronology of Innovations

Appendix: *From Stone to Cement*   207

Section III: The Cave and the Quarry from 1990 On

1. Howe Caverns from 1990 On   210
   Pushing Beyond the "Mystery Passage" for More Cave

2. Rebirth of the Howes Cave Quarry   225
   Bold Plans Connect Present with Historic Past

3. Sold!   236
   A Family Says Goodbye

4.	A Word about White-Nose Syndrome	241
	Howe's Cave "Ground Zero" for Deadly Fungus	
5.	New Owners, New Outlook	244
	Adventure, an Animatron, Naked Tours, and More	
6.	New Owners, New Outlook II	257
	What Almost Was	
7.	The Cave House Museum of Mining and Geology	264
	A Love Affair with Underground Empires	
	Epilogue	282
	Connecting the Cave Community	

Section IV: Related Tales from the Cave Country

1.	The Gebhards of Schoharie	288
	Father and Son Pave Way for Earliest Studies of Geology, Natural History	
2.	1889 Photos Offer Rare Look Underground at Old Howe's Cave	297
	Images by Noted Adirondacks Photographer Capture Lost Underground	
3.	Here She Caves . . . Miss America	302
	And One Man's Attempt to Open a Schoharie Cave	
4.	From "Nameless" to "Secret"	307
	Flashlight Company Shines a Light on the "Lesser Caves" of Schoharie County	
5.	Blenheim Monster Serpent Mystery Unsolved	311
	Beast Slithered from Its Mountain Cave	

Published Sources	314
Index to *Underground Empire*'s Cast of Characters	315
The Explorers, Entrepreneurs, Promoters, Scoundrels, Scientists, Innovators, and Aficionados	

Foreword

We always thought we knew a lot about Howe Caverns and the other geological marvels of Schoharie County's "Cave Country." After all, we have been there so many times, including New Year's Eve, 1999. (Remember the "Y2Cave" event?) But we have learned so much more from *Underground Empires*. Dana Cudmore has been able to put together what might be called a "synthesis" of its history. And the history of the famous cave and the geology of the surrounding area is of some importance to us as Catskills' residents. After all, Howe Caverns remains our region's single most successful tourist attraction and, together with Secret Caverns, one of the area's most vital economic engines. Hand-in-hand with the story of the caves is the story of the region's cement and stone industry. We really should be familiar with this history—it's important. At one point, the Howes Cave cement quarry was Schoharie County's largest employer. Stone and cement from the mine and quarry have gone to important projects across the state.

One fact that struck us both early in this book was just how long Howe Caverns, Schoharie County's first commercial cave, has been operating. Tours began there in 1842 and have continued, off-and-on, ever since. The region's cement industry began not long after, extracting the same limestones that compose the caves. That's a lot of time, and that makes for a lot of history.

Cudmore has written a book not solely about the caves' history, but about all the history attached to the caves—the multitude of personalities and the cultural and economic tides that have swept across the Leatherstocking Country and its greatest natural wonders.

It all began in 1842 when local farmer Lester Howe went looking for caves—and found a gem. Lester should have been one of countless nineteenth-century farmers who have become lost to time. Cudmore speculates on Howe's motivations and then describes the man's first enthusiastic days of cave exploration. And "enthusiastic" is exactly the right word for

Foreword

Lester Howe. He rearranged his life in opening the cave to tourists and building a hotel (the first Cave House) where they could stay in comfort. Lester was not a front-office sort of person. He led the tours as often as he could. And he was, apparently, quite the ham; he was entertainer as well as guide, and good at both.

This was a time when Americans were flocking to immerse themselves in natural history. Other great commercial caves—Luray Caverns in Virginia and Mammoth Cave in Kentucky—were doing very well at that time. The Catskill Mountain House Hotel brought tourists to enjoy the views from the Catskill Front, and the Trenton Falls Hotel brought them to New York State's Trenton Falls. The Hudson River School of landscape painters was at the height of its popularity. It was a national celebration of America's natural environment.

Lester Howe's tenure at the caverns would last only until the 1870s, and then his enthusiasm would get the worst of him. His passion for expanding the caverns and the hotel was not matched by the very best business sense. That left him financially overextended and vulnerable to business predators. That soon brought in the next leading personality in this story— Joseph Ramsey, who seems to have had far less enthusiasm for the cave than in the money that could be raked in from it and from his nearby cement works. History remembers Ramsey as a cutthroat businessman who won control of the cave through a business deal that, let us be polite, Howe did not *fully* understand. Howe left, and Ramsey took over.

Meanwhile, a historic irony was underway. The very same limestone that made up Howe Caverns was also well suited for construction purposes and for the production of natural cement. Back then it was the most advanced way of producing high-quality cement. A very sizable limestone mine began beneath the adjoining property, and a surface quarry began the extraction of building stone. Both the quarry and the mine were advancing toward the cave itself. A long-term stress was being generated by the close association of the economics of tourism and of industry. We were shocked to learn that fully 300 feet of the historic cave were swallowed up by the quarry—gone, forever.

The early twentieth century saw a regional economy that had nearly dropped its tourism foundations and become thoroughly industrial. Howe's Cave was virtually closed from 1900 to 1925. Elsewhere, the Catskill Mountain House hotel was fading and the Trenton Falls Hotel closed. America had seemingly lost its interest in Nature.

But new, modern engineering advances led, eventually, to the next generation of people to run Lester Howe's cave. John Mosner and the Sagendorf family were developers in the full modern sense. They envisioned a cavern that could be entered by elevator, lit by electricity, and toured in comfort, and they made it happen. The modern Howe Caverns made its appearance in 1929, and today's twenty-first-century cave is relatively little changed in its essentials from what was there ninety years ago.

Another startling fact presented in this book is that there are at least 150 other caves in Schoharie County. In the months and years that followed the reopening of Howe Caverns, three of these other caves opened as competing tourist attractions. These were smaller caves, and none had the budgets to match Howe's. They generally made the best of that by advertising themselves as being "more natural" and less commercialized than Howe Caverns. Of these, only Secret Caverns remains open today.

Howe Caverns recently changed hands again, and the current owners have brought a new vision to it. They see it as an adventure theme park as well as a cave tour. There is a rope course, a zip line, and a rock-climbing tower. For the most adventurous, there is the naked cave tour, which is just what it sounds like and has always sold out. All this seems to be working; paid tours of the cave have been growing.

The old Cave House Hotel, long derelict, has become the Museum of Mining and Geology. It has been slowly evolving and is open most summer weekends. An annual autumn gem and mineral show is a highly recommended event.

It is ironic that the same industry that nearly destroyed Lester Howe's cave in the nineteenth century has revitalized it in the twenty-first, along with reclaiming and making use of the long-abandoned limestone quarry

Foreword

and the old hotel. It is a rare successful nexus of industry and tourism, commercial success and preservation.

Underground Empires is a rise, fall, rise, fall, and rise again story filled with colorful characters, high adventure, a little romance even, and some tantalizing mysteries. Where, for example, is Lester Howe's legendary Garden of Eden Cave, which he claimed was "bigger and better" than Howe Caverns? For that information, read on …

Robert and Johanna Titus

June 2021

Robert and Johanna Titus are the authors of *The Catskills: A Geologic Guide, The Catskills in the Ice Age, The Hudson Valley in the Ice Age*, and other books.

On the following two pages

A map of the Cobleskill Plateau, showing some of the key locations in *Underground Empires*, by karst hydrologist and caver Paul Rubin. In addition to caves in the area, it also shows numerous sinkholes—indicated by dots—and underground drainage patterns, shown as dotted lines. Drainage flows generally due south toward Cobleskill Creek. You can also follow the "Finger of Geology" from McFail's Cave through the "missing link" to Howe Caverns and beyond to Schoharie and VanVliet's Cave (not indicated). You can also see the path of the historic Albany and Susquehanna Railroad, now the D&H.

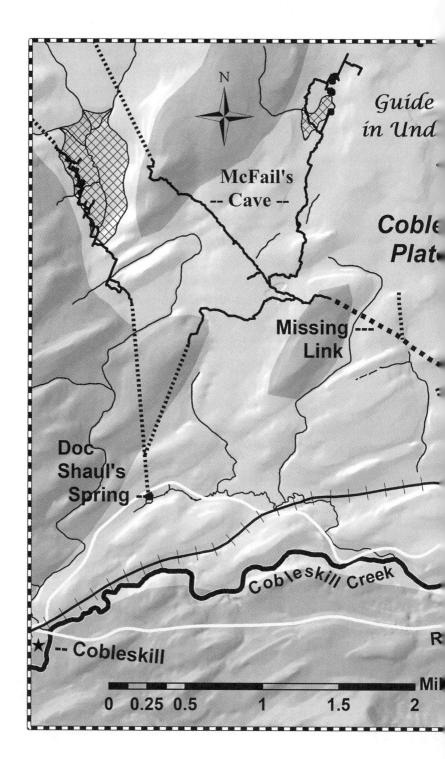

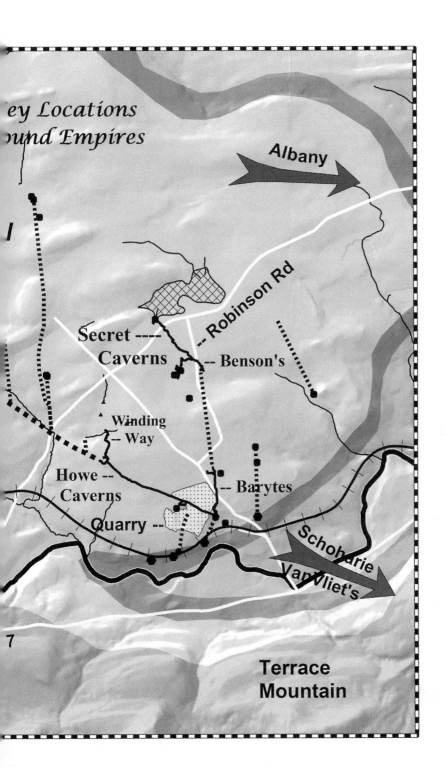

☞ Edison's Electric Light in the Cave, with a Magic Lantern will be an additional attraction the present season.

An artist's sketch of the Cave House and expansive, elegant Pavilion Hotel from the owners' 24-page brochure, "A Summer Home. The Pavilion Hotel, Howes Cave, NY." The lengthy travel guide from 1889 asked and answered the question: "Where Shall We Go." It is available for download on the Library of Congress Web site.

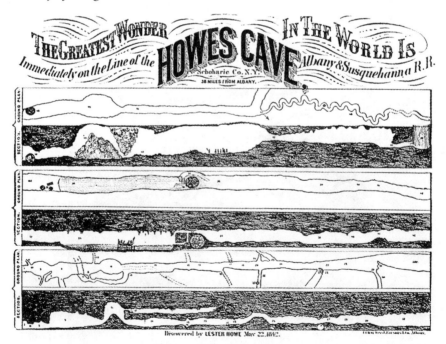

From the 1889 "A Summer Home" brochure, an artist's sketch of "Crystal Lake—Pulpit Rock," now the Lake of Venus and the Cathedral Pipe Organ. The artist has about doubled the size of these picturesque formations.

At left: A map from "A Description of Howe's Cave," published in 1865 by Weed and Parsons of Albany. Note the passengers in the underground boat.

Noted Adirondacks photographer S.R. Stoddard toured Howe's Cave in 1889 to capture several images for publication and posterity. Above, a band performs in the caverns' "Music Room," presumably named for its concert hall acoustics. Photo courtesy Library of Congress.

Weddings have been performed underground in Howe's Cave since as early as 1852. The first were performed in the "Bridal Chamber," a natural alcove not far from the cave entrance. The ceremony above, possibly staged, was captured by photographer S.R. Stoddard in 1889. Photo courtesy Library of Congress.

Acknowledgments

This work is a compilation of the author's own research and the numerous articles about Howe Caverns, Secret Caverns, and the caves of Schoharie County, New York. Together these records document a dramatic history that has all the elements of a Shakespearean play. In turns this history is exhilarating, funny, infuriating, frightening, tragic, and inspiring. It has many memorable characters, heroes, and villains—eccentrics and men of science, farmers and shrewd rail tycoons, rivals, intrepid explorers—and even some underground romance.

The sources for this history include numerous personal accounts and published recollections of witnesses and those that helped shape the events described. Many of these accounts are from more than a century ago. Details have also been pulled from countless books, newspapers and magazine articles, brochures (which often tell a story), personal correspondence, and specialty publications such as industry newsletters, etc. This book is intended to be an entertaining and fulsome look at nearly 200 years of cave and community history. Yet, it is not a complete account of the story of the cave found by Lester Howe and of what followed his discovery. If that story were possible to compile, the complete work would fill volumes.

Most acknowledgments fall short when trying to recognize all those individuals that deserved to be recognized. I suspect that is the case here as well, but I will attempt to fully credit individuals who contributed over the span of some thirty years, before and since the 1990 publication of *The Remarkable Howe Caverns Story*, and I further hope to acknowledge all of those who contributed to the other sections of this book.

I am deeply indebted to the following: **Emil**, **Michael**, and **Sam Galasso**, for their ongoing involvement in Howe Caverns, the Howes Cave Quarry, and Cave House Museum of Mining and Geology; **Clemens McGiver**, for his vision and twenty-plus-year commitment to the adaptive reuse of the long-abandoned quarry and its unique features; **Robert Holt**, a former

Acknowledgments

general manager at Howe Caverns who "filled in" a lot of the missing pieces, provided access to the caverns' historical archives, and provided many of the rare photographs; cavers and old friends **Robert Addis** and **Chuck Porter**, for background, research, great photos and tales; karst hydrologist/caver **Paul Rubin**, for access to his exciting work that one day may discover the "Missing Link" at the cave; **Nancy Sagendorf** and **John Sagendorf**, for sharing insights on the family's long ownership of the cave and its sale; **Len Berdan**, former member of the Howe Caverns board of directors, who contributed details on the sale of the cave; **Helena Ackley**, for details and photos of the Robinson family's involvement in the area's caves; **Harrison Terk**, another former caverns' manager, who talked about the sale and gave me permission in 1989 to explore beyond the "fat man's misery" section; former cave employees **Horace Rickard** and **George Smith**, for reminiscing about the early days; cavers **Luke Mazza**, **Gordon Smith**, and **John Dunham**, editor of *The Northeastern Caver*, the best regional caving publication in the country; caver **Thom Engel**, author of *To Rival Mammoth Cave: Howe's Cave before it was Howe Caverns*; **Rich Nethaway**, for help with the Cave House Museum history; **Roger H. Mallery**, the second-generation owner of Secret Caverns; **John F. Meenehen**, who helped research the methods of early cave photography, and **Timothy Weidner**, director of the Chapman Museum of Glens Falls, for the same; **Warren Howe**, for his assistance with the Howe family genealogy; **Jim Poole**, publisher of the Cobleskill *Times-Journal*, for access to newspaper editions covering the years 1926 to 1931; **Helene Farrell**, former curator of the Schoharie County Historical Society; Director **Kimberly Zimmer** of the Cobleskill Community Library, and **Timothy Holmes**, the former librarian; **Stephanie McClinton** of the Kansas City Public Library, for records on the Howe family in Jefferson City, Missouri; **Paul D. Bonvoe**, with the SUNY Cobleskill Library; Cave House Museum trustee and NYS Geologist **Paul Griggs**, for his expertise in geology and quarrying; NYS geologist **Dr. William Kelly**, for his early advocacy of the Cave House Museum and help with its accreditation; **Anna Bautochka, Bob DeRuvo, Fred Boreali, Ken Braman, Julia Fullerton Irwin, Shirley LaBadia, Lavina Emery Mulbury,**

John Pangman, Tony Spenello Jr., Dennis Tillison, Richard Wick Jr., Forest Wollaber Jr., and **T.L. Wright,** all of whom have fond memories of growing up in the Howes Cave hamlet and/or working at the quarry; **Anne Hendrix,** for recollections of her uncle, Colonel Rew, and longtime family friend, Arthur Van Voris; **John Murray Sr.,** a former guide and member of the board who tells an illustrative cave/quarry tale; Esperance historian **Ken Jones,** for a historical tidbit or two; any former employee who can still give a tour decades later; and to all who in any way over the years have made contributions to the body of literature on Howe's Cave.

A special thanks to the late **Ben Guenther**, a friend and caver, who brought me into the early efforts to establish the Cave House Museum of Mining and Geology.

Note: These acknowledgements are intended to recognize those who have contributed to any or all of the full four sections of this book. My sincere apologies to anyone who should have been included but was not.

Preface

This book explores the wonder and drama embedded in the history of the caves in rural Schoharie County, New York, and the effect the caves have had on their discoverers, owners, and millions of visitors from around the world. It offers a look at the people and communities that developed around the caves, as well as the rugged industry that was built on that unique natural footing. Dana Cudmore, the author, grew up in the middle of this "cave country."

This edition includes an updated and expanded version of the author's 1990 book *The Remarkable Howe Caverns Story*. Previously undocumented details taken from years of accumulated research have been added throughout this new, 175-year-plus history, plus a great deal of history has been made at Howe Caverns and its related environs since *The Remarkable Howe Caverns Story* was first published. *Underground Empires* updates for the reader the remarkable changes that have taken place since 1990, including the sale of the cave, the reopening of the quarry, and the creation of a museum of mining and geology dedicated to this unique historical and industrial setting.

The fascinating story and descriptions of Secret Caverns have also been greatly expanded, and additional photos—many rare and from personal collections—have been added.

Also featured is the author's history of the Howes Cave stone and cement quarry and the hardscrabble community that grew up around it, which was originally self-published in 2005 as *Unearthing Howes Cave*. That history has also been updated to include additional research and newly added photos.

This book is divided into four sections. The first includes the expanded *Remarkable Howe Caverns Story*. This is the history of the caves through 1990. (There are a few "time jumps" to make use of context and to ease understanding.) Section II includes historical content on the cement industry in Howes Cave, expanded from the content in *Unearthing Howes Cave*.

The third section covers new developments since 1990 at Howes Cave and the quarry, including the new ownership, new attractions at the cave, and the rehabilitation of the Howe family's Cave House as a museum.

In Section IV there are "bonus" chapters on five related historical topics. The first describes the work of John Gebhard and his son, John Jr., both pioneering geologists in the Schoharie Valley. The second describes the challenges faced by renowned Adirondacks photographer/naturalist S.R. Stoddard, who took photos in Howe's Cave in 1889. His remarkable photos are used throughout this book. All the glamour and pageantry of a 1958 visit by Miss America to a tiny Schoharie cave is recalled in chapter three. The fourth chapter describes how a Hoboken battery and flashlight company partnered with a Cobleskill hardware store owner to promote the "lesser caves of Schoharie County" and pave the way for the opening of Secret Caverns. The final chapter documents the obscure story of a mysterious "monster serpent" that slithered down a Town of Blenheim hillside in the early 1800s. Photos are included throughout Section IV.

The Title of This Book

The title was chosen as homage to Clay Perry (1887–1961), a Pittsfield, Massachusetts, journalist and caver who wrote about the caves of New England and elsewhere in the northeastern United States. His 1948 book *Underground Empire: Wonders and Tales of New York Caves* was read time and again by the author and is often referenced here. Coincidentally, both Perry and the author worked as journalists.

Perry is credited with coining the term "spelunker" in the 1940s (from Latin *spelunca*, which in turn derives from the Greek *spelynx*, both meaning "cave"). The word is rarely used by underground explorers, who prefer "caver."

Perry was influential on the sport of caving in other respects as well. On December 1, 1940, a group of twenty-four cavers under Perry's leadership met in Pettibone Falls Cave, Massachusetts. There, they ratified the proposed constitution for a National Speleological Society (NSS), a nonprofit organization to "advance the exploration, conservation, study, and understanding of caves in the United States."

Today, there are 250 grottoes, or chapters, of the NSS.

About the Author

Author Dana Cudmore outside the old entrance to Howe's Cave on the grounds of The Cave House Museum.

Photograph by Beth Moore.

Dana Cudmore of Cobleskill is a retired communications professional who worked as a newspaper reporter and editor as well as a public relations director in the New York State university system. After running his own agency, Media Services, for a dozen years, he worked as an external affairs officer for a federal disaster response agency before retiring.

Among other distinctions, he was a member of the original board of directors of the Cave House Museum of Mining and Geology, created in Howes Cave in 2003.

Cudmore put himself through college working summers at Howe Caverns, where he developed an interest in its history and folklore. As a young man he explored many of the region's other, undeveloped caves, as well as caves in Missouri, Pennsylvania, Vermont, Virginia, and West Virginia.

He writes: *Growing up in nearby Central Bridge, I was one of the neighborhood kids that rode our bikes along the D&H Railroad tracks to the hamlet of Howes Cave. There wasn't much to it in the late '60s–early 1970s. The quarry was operating only at a limited capacity; there were a few buildings on a "main street" and a lot of boarded-up barns and sheet-metal sheds. Everything was grey from the cement dust. Along the road there were several wired-up, dangerous-looking entrances underground to caves or the* nineteenth-century mine. "Do Not Enter" and "Danger—Stay Out" signs were everywhere. I stayed out.

Like so many other locals, I never visited Howe Caverns until I took a summer job there in 1970. The annual summer workforce consisted almost entirely of teenagers from the four nearest central school systems. I made a lot of friends and had great summers there. Giving four tours a day, I also learned a lot about people from all over the country, and yes, the world.

The 52° "other world" of the cave offered a bit of thrill, and I was fascinated by its mysterious dark passages—some unseen and unavailable to the public. I wanted to explore and to know all I could. I read and reread sections of Perry's *Underground Empire* many times. My interest in the symbiotic Howes Cave quarry and its community took longer to come into focus, as evident in the two different publication dates of my original works.

So here we are. I hope you'll enjoy this uniquely American history of this truly one-of-a-kind upstate New York setting.

A Note about Howe, Howes, Howe's

Howe's Cave is the original name for the underground wonder, from its discoverer, Lester Howe. Around 1890 the owners of the adjacent cement quarry dropped the apostrophe, and that convention is used for the community and on the highway signs that lead there. Howe Caverns was the corporation created to open the cave as a modern, well-lit tourist attraction in 1929, and that name is still used today. Every effort was made to use the name appropriately throughout the text.

SECTION I

The Remarkable Howe Caverns Story

Introduction

A History with Parallels to the Growth of a Nation

"Certain places extend the spell of their glory to the uttermost parts of the earth, and a long procession of fascinated pilgrims ever wends its way to such world-renowned shrines. One of these is Howe Caverns, made glorious by the magic of unknown ages."

— from *The Story of Howe Caverns*, 1936

The story of Howe Caverns in upstate New York closely parallels that of nineteenth- and early-twentieth-century America. Discovered twenty years before the start of the Civil War, the cave's history mirrors our nation's transformation from a farm-based to a factory economy. It was a time when great fortunes were made by leaders of this industrial revolution, and an emerging modern society was eager to exert its control over Mother Nature.

There are more than 17,000 caves in the United States, and of those about 80 are open to the public, according to the National Caves Association. Mammoth Cave in Kentucky, Carlsbad Caverns in New Mexico, Luray Caverns in Virginia's Shenandoah Valley, and Howe Caverns in central New York are easily the most heavily visited. Howe's Cave, as it was first known, was open for tourists in 1842. It was America's third great commercial cave. Virginia's Weyer's Cave—now Grand Caverns—took in paying customers as early as 1806. Mammoth Cave opened in the mid-1830s; Luray in 1875. Carlsbad, regarded by many as America's most beautiful cave, didn't become a show cave until the mid-1920s. That these attractions continue to draw hundreds of thousands of paying visitors each year is a testament to their natural appeal, their locations, and their owners' sharp business sense. Profitability is relevant—and impressive—in the case of both Howe and Luray caverns, which are privately owned. (Mammoth Cave and Carlsbad

Introduction

Caverns are run by the National Parks Service.) In 1927, when plans were being made for the revitalization of Howe's Cave, its developers astutely noted that fully one-quarter of the population of the United States lived within only one day's drive of the cave.

WELCOME TO HOWE'S CAVE—A rare photo of Lester Howe, standing at the stone-fortified entrance to his cave, in about 1870. The image was part of a series taken for stereographic viewing. Only a few images remain.

What makes Howe Caverns unique? Though its size is dwarfed by the more than 400 miles in the nation's longest cave, Mammoth Cave, and its decorative rock formations are less abundant that Carlsbad or Luray, Howe Caverns boasts what is probably the most remarkable history of any cave in America.

Taking the one-hour-fifteen-minute Howe Caverns tour today, one has a difficult time imagining the way it once was—a muddy, strenuous, all-day expedition lit by crude oil lanterns, and yet gladly patronized by the curious, well-to-do vacationers of the mid-1800s. It is equally difficult to imagine the gargantuan efforts undertaken over the years to develop the caverns and make it a comfortable trip for the touring public, as they see it today.

The full story of Howe Caverns—and its neighbor, Secret Caverns—is not widely known and undeservedly so. In the more than 175 years in which Howe Caverns has been a show cave, it has had a profound effect on the lives of the explorers, developers, and tourists associated with it, as well as on the small community in which the cave was found. Lives have been lost, fortunes have been created and destroyed, and bitter rivalries have developed while competing for the tourist dollar. A modern-day folktale has even evolved, describing a legendary and hidden cavern, bigger and better than Howe Caverns itself.

It is this fascinating human drama behind the discovery, exploitation, demise, and resurrection of Howe Caverns that earns the cave a unique position in the history of American caves.

And we'll see in later chapters how Secret Caverns and Howe Caverns are inextricably linked, as is the Howes Cave stone and cement quarry. Although Secret Caverns has a much shorter history to share (it was first fully explored in the late 1920s), that story is also compelling … and perhaps a little eccentric.

1

Earliest Records

A Short History of the Schoharie Valley;
Indian Legends and a Revolutionary War Hideout;
The Story behind Lester Howe's 1842 Discovery

Howe Caverns winds for nearly a mile beneath the rolling hills of rural Schoharie County in upstate New York's central "Leatherstocking Region," a name derived from the protective leather leggings worn by colonial-era pioneers. The phrase was made famous in the classic works *The Last Mohican* and *The Deerslayer*, by America's first novelist, James Fenimore Cooper. To the east is the Hudson River Valley; to the south lie the foothills of the Catskill Mountains; on the north and west is the historic Mohawk Valley. The Leatherstocking Region is one of the most picturesque areas of New York State. In terms of geology, Schoharie County is the northeasternmost point of a great "cave belt," the geological remnant of an ancient ocean bed that arcs south along the Appalachian Mountains through Pennsylvania, the Virginias, Kentucky, Tennessee, and westward to the coast.

Howe Caverns' tour guides (and a new, animatronic Lester Howe) tell the conventional story of the cave's discovery in 1842 by Howe, a Schoharie County farmer. But the actual discovery may have taken place more than a century before, and the Schoharie Valley was rich in history when Lester Howe first entered the cave that still bears his name.

One hundred and fifty families of refugees from the Palatinate area of southwestern Germany settled the area in the early spring of 1713. To reach the Schoharie Valley, it took a two-day trek from the Hudson River settlement at Albany along a foot trail over the high Helderbergs. The resident Native Americans—a tribe of about two thousand mixed clans recognized by the bear insignia of the Mohawks—had settled the area not many years prior. The valley through which the Schoharie Creek flows was a paradise for the first European settlers, who located in seven camps, or "dorfs," along the riverbanks. (Yes, there were seven dorfs. It's the German word for "village" or "town.") Though not without hardship, the area was well established with frontier settlements at the outbreak of British and Indian hostilities in the late 1700s. The valley's precarious position as New York's westernmost border in the Mohawk Valley led to much bloodshed.

The area's unique geology, which includes numerous caves in the northern half of the county, is noted in the earliest historical records. (There are about 200 known caves in Schoharie County and about the same number in adjacent Albany County.) The underground even found its way into the German names for the settlements. What is today the village of Schoharie was at first Brunnen Dorf, or "Fountain Town." The fountain is a still-flowing spring that bursts from an unknown underground source near the Palatine House Museum, appropriately located on Spring Street.

During the Revolutionary War, the Schoharie Committee of Safety—a six-member organization responsible for wartime provisions—held secret meetings in what was called the "committee hole," near Middleburgh. In the 1845 *History of Schoharie County and Border Wars*, author Jeptha R. Simms quotes from the papers of Judge Peter Swart (1752–1829): "In August, 1777 . . . I was one of the six councillors [sic] that went from the stone house across Schoharie Creek into the woods to a cave to consult what measures to adopt—secrecy at that time was the best policy." Simms describes the cave as being on the opposite side of the river from Middleburgh, in the ravine between the mountains. The stone house Swart describes was known as Fort Defyance (sic).

There is also speculation that Ball's Cave, on the hills high above the village of Schoharie (the county seat) was once a Tory rendezvous. Peter Ball, the Barton Hill landowner for whom the cave was named, was chair-

Earliest Records

man of the Committee of Safety. (Ball's Cave was Gebhard's Cave for a time in the mid-1800s, and is now Gage's Cave and a part of the James Gage Karst Nature Preserve of the National Speleological Society.)

Prior to the arrival of the German settlers, many believe the local Mohawk tribes knew of what is now Howe Caverns and called it "Otsgaragee," or "Cave of the Great Galleries." In the historical records there is some disagreement as to this translation, but if accurate it would suggest that the tribe explored deep into the cave's interior. Archaeologist E. George Squier (1821–1888) of Albany visited the cave in 1842 and reported, "Human bones as well as pieces of charcoal, encrusted with a solid coating of carbonate of lime of two to three inches in thickness have been found at the distance of more than a mile from the entrance." This is an unverified secondhand report; a second translation, "Great Valley Cave," may be more accurate. A third translation is "Hemp Hill," a Mohawk reference to the area around East Cobleskill and the cave's entrance.

The first white man to enter the cave may have done so in the late 1770s. Perhaps peddler Jonathan Schmul had been calling on families settled in the area east of Kobel's Kill—"Kobel's Creek," today's Cobleskill—when he sought refuge from a Mohawk hunting party by hiding at the entrance to the cave. Schmul's story may have been recorded firsthand by the colorful "Forest Parson," the Rev. John Peter Resig, and first published in Germany. Resig founded a German Evangelical Church on the Schoharie and kept a diary that stands as a great historical record of the period. According to the diary, the peddler Schmul was "well informed on matters pertaining to the strained relations" between the Indians, British, and settlers, and took Pastor Resig into his confidence. Upon being asked where he lived, Schmul answered: "I have revealed that to no one, but since you are a minister and keep the secrets of the confessional, I'll tell you. Ten miles west is a creek named after the German Kobel, that is Kobelscreek. There I found a cave when the Indians were after me. That's my home. But be mum about this. Should war break out, then flee to this cave and you will be safe." Later, Resig was called upon to visit a sick woman who lay by candlelight in the dimly lit cave.

The diary also describes Resig's Revolutionary War exploits. He once left the cave and hastened to the scene of the conflict at Oriskany, arriving just as Chief Brant pointed out General Herkimer to the Indians, who made a dash for him." The heroic pastor sprang to the general's side and seized a battle ax in his defense.

The diary provides a great tale, but there are a number of problems with this account, which for several decades was included in the Howe Caverns Corporation's *Story of Howe Caverns*, a pocket-sized, hardcover souvenir. It started with Pastor Resig's story being told in *Der Waldpfarrer am Schoharie*, published in Germany by Dr. Friedrich Mayer in 1911 as a work of fiction. *Der Waldpfarrer* translates to "The Forest Parson," and the subtitle of the book references it as a look at German life in America in the eighteenth century. Howe Caverns Incorporated may have picked up the tale from its 1931 English translation, *Fifty Years in the Wilderness*, by August William Reinhard, for Wetzel Publishing, Los Angeles. Further, there is no mention of a Pastor Resig in either of the nineteenth-century histories of Schoharie County, which otherwise provide exhaustive lists of all local clergy in that period.

It follows then that Schmul and Resig vanish quite suddenly from the historical records, as did the natives of the Schoharie Valley, who fled the area with their Tory counterparts at the end of the Revolution. By the time Lester Howe settled in the valley east of Cobleskill, there was little known or remembered of Otsgaragee. The location had been lost to history. But there was talk of a mysterious "blowing rock"—a strange rock ledge from which a cool breeze emanated on even the hottest days—"so cold and strong, that in summer it chilled the hunter as he passed near it," according to an April 1857 account in *The National Magazine*. "No person ventured to remove the underbrush and rubbish that obscured the entrance, lest some hobgoblin, wild beast, or 'airy creature of the elements,' should pounce upon him as its legal prey."

There was great, yet primitive, interest in the natural sciences in the early 1800s. In the Schoharie Valley a spontaneous wave of cave exploration and interest in the natural sciences took place during the years

Earliest Records

1820–1850. Speculation suggests that this can be traced, in part, to the cave explorations in the frontier country of Kentucky, much publicized by the newspapers of the day. Shortly after the War of 1812, the mummified remains of prehistoric native explorers were found in Mammoth Cave. Soon thereafter a profitable attraction was established to meet the demands of a curious paying public.

A small group of naturalists and amateur geologists, notably Schoharie attorney John Gebhard and his illegitimate son, John Jr., discovered and explored local caves, identified and amassed a huge collection of fossils, rocks, and minerals, and documented numerous geological characteristics of Schoharie County.

The caves were plundered for their unique mineral formations. In 1831, Gebhard's Cave (previously Ball's Cave and now Gage's Cave), was explored, mapped, and made famous by the newspaper writers of the day, notably E.F. Yates, a correspondent for the *New York Commercial Advertiser*. In 1835, John Sr. published "On the Geology and Mineralogy of Schoharie" in the *American Journal of Science*.

In 1836, John Jr. was appointed as a district assistant to the New York Geological Survey, a huge undertaking to document the state's vast natural resources. His team's assignment included Schoharie County; the findings were published in the survey's first report, released in 1843. (A chapter in Section IV is devoted to this pioneering father-son duo and their work.)

T.N. McFail, a professor at the seminary school in Carlisle in Schoharie County's northern cave country, also explored many of the area's caves. In 1853, McFail died in a fall while climbing from a cave that today bears his name. One old-timer with an interest in cave explorations wrote:

> The last climb was too much for him. Professor McFail, I might explain, was exceptionally large. It was finally decided, after several unsuccessful attempts, to tie a loop in a second rope into which he could place one foot. He was able to pull himself up, and his companions were to keep the second rope tight so that he could rest . . . In this he slowly made his way up until his hat showed above the

Lester Howe, 1810–1888.

edge. Then, all of a sudden, the weight was gone from the rope and a dull thud came from the bottom.

According to native lore, his fine gold watch and silk scarf were never recovered.

So, it seems quite likely Howe was familiar with these adventures and possibly took part in some. His family, including three brothers and two sisters, had settled in the neighboring Otsego County community of Worcester at the turn of the century. Howe's grandfather, Elijah, as a military man and town tax collector, was pitted against his neighbors in Belchertown, Massachusetts, during the anti-rent rebellion led by Daniel Shay. It may have been that public opinion forced the Howes to leave their home in Belchertown for upstate New York.

Lester Howe was born January 7, 1810, in Decatur in Otsego County, the second of six children born to Ezekiel and Nancy Howe.

In 1842, at the time of his celebrated discovery, Howe and his wife, Lucinda (Rowley) Howe, and their three infant children—Huldah, Harriet, and Halsey John—were a young farm family, having recently settled property in East Cobleskill adjacent to the caverns' hidden entrance. The property had been purchased by Howe's father from Lucinda's brother, Julius. The farm home faced north from the southern slope of a mile-wide valley forged by an ice age glacier eons ago and through which the Cobleskill Creek wound its way.

The farm was equidistant from the villages of Cobleskill and Schoharie, the county seat. Hops, used medicinally and for flavoring beer, was the area's most important cash crop of the early to mid-1800s. Like their neighbors, the

Earliest Records

Howes probably had a hop house for storing their crop and kept a few cows for milk and butter.

There is no doubt that Howe found alluring the story of the strange local phenomenon, "the blowing rock." Reports placed its location just north of the "Kobels Kill" and ten miles east of the Schoharie Creek, on or near his property.

There are several different accounts of the caverns' discovery. The most often told, simplified for the touring public, is that Howe found the cave on his property by accident. His dairy herd milled about near the cave's hidden entrance to feel the cool air coming from below. (There is probably some element of truth to this; modern-day bovines have been seen enjoying the cool breeze from tiny Young's Cave, just north of Howe.) In particular, for many years a cow named "Millicent" was given credit for helping with the discovery.

Howe's own story even changes over the years. An 1880 promotional account provides the following:

> Howe's cave was discovered by Lester Howe, whose name it bears, a Schoharie County farmer, who still lives in the vicinity. Howe's own account of the discovery is as follows: While rabbit hunting one day in the fields over this cave, he fell into a "sink hole." Relating this adventure to a scientific man he was informed that sink-holes mark the course of extinct or existing caves. He had also noticed that on hot summer days, the cows that were pastured in a field at the present mouth of the cave, were accustomed to huddle together to a certain spot that was in reality less shaded than other parts of the field. But the most singular circumstance was the fact that the temperature in this particular portion was very much cooler than the general temperature. Howe now began to believe that his farm contained a cavern. On the 22d of May 1842, he and some friends were fox hunting. The fox secreted himself in the face of the hill where the entrance to the cave is today. In digging the animal out, the cave was discovered.

Whatever his reasons, throughout 1841 and the early months of the following year, Howe was actively looking for a cave or caves. The historic records are agreed on the date—on May 22, 1842, Howe entered the "blowing rock" after discovering its location on the adjacent property of Henry Wetsel. His neighbor probably accompanied him.

Much to the chagrin of his wife, Howe, "with commendable curiosity," returned to his discovery day after day, often with Wetsel. There is no firsthand account of Howe's first explorations. Likely, he and Wetsel ventured a little farther into the cave on each trip and emerged wet, muddy, and exultant with the thrill of their discoveries. A piece of tin

The printer's woodcut of the original, natural entrance to Howe's Cave, from the 1843 Geology of New York. There appears to be a crude wooden gate above the entrance and some stone work and stairs as well.

was hammered into a lamp to burn whale oil, creating for the explorers an easily carried and reliable source of light. Shaped like a funnel, this meager light featured a unique collar, fashioned to protect the wick from the damp caverns' elements.

In its natural state, Howe's Cave had a grandeur all its own. It is totally unlike almost all other northeastern caves, which require a considerable amount of physical exertion—crawling, climbing, and squeezing through tight, tortuous passages, some filled with 42° water. Howe's Cave was an easy trip, relatively speaking. The cave consists mostly of walking passages and large rooms fifteen to thirty feet wide and ten to sixty feet high. A cold, crystal stream flows throughout most of the cave's one-and-a-quarter-mile course.

But caves exist in utter darkness, entirely remote from any hint of daylight. The gray limestone walls swallow even the most powerful of lights. Led by a flickering oil lamp not much brighter than a candle, Howe's and Wetsel's explorations lasted eight, ten, and twelve hours, and possibly longer. The low, narrow entranceway led to a wide, airy chamber, later named the "Lecture Room." All about the two explorers were broken fragments of the caverns' ceiling—tremendous blocks of limestone rubble—which in places nearly filled the passageway. The danger of a cave-in is always more imagined that real, yet the effect of walking over, around, or through rocks that have fallen from the ceiling eons ago is still frightening. A visitor in 1851 wrote, "Never in my life have I seen such rocks—sometimes piles of one above another, and seemingly ready to fall at a touch, and now grand boulders of immense size pushing out from the walls and almost 40 feet above our heads."

As the explorers left the Lecture Room, the light of their lamps was lost in the high-domed ceiling, which shot up 60 feet into impenetrable darkness. The passage split into parallel paths, two grandiose tunnels each about 600 feet long, about 10 feet wide and between 10 and 30 feet in height. The explorers discovered that the passages rejoined, and they then entered another large chamber, later named the "Giant's Chapel." Howe and Wetsel would have been in the cave more than two hours at this point and had traveled only about 1,000 feet. They rested in the chapel, a large room about 40 feet wide by 40 feet long by 50 feet high.

It is at this point that an underground brook crosses the main cave passage, and Howe and Wetsel continued upstream. A later explorer would write romantically:

> Our path . . . was by the side of a little stream of clear, cold water, and before we had gone half a mile we could hear its babble and its dash, which seemed to gather volume as it approached and broke at length upon our ears like the voices of many waters. We crossed the stream I should think a hundred times. Sometimes it was a chasm below us, then spread out across our path, then babbling merrily at our side—its far-off rush the while sounded as if cataracts and waterfalls were plunging into some abyss together. A stream of water underground whose source no one knows!

Beyond the Giant's Chapel were several small side passages heading off the main passage to entice the explorers. Poking and probing the small openings, they discovered another curiosity and later named this section "Cataract Hall." An early visitor described it:

> There is a small opening in one side which extends far into the mountains. You put your ear down and listen there and you are astonished and start back. You listen again, and you hear the sound of a cataract pouring over rocks, and you listen till you are filled with awe. There far off in the mountain, where man never was or ever can be, where the Creator's hand and power alone has been, goes down some grand waterfall, just as in the beginning. He made it to go, and not the sight of which, but only the sound thereof man is permitted to enjoy.

Howe and Wetsel continued for only about another 600 feet before being stopped at the head of a crystal-clear underground lake. They were unable to cross, but could see that the cave continued beyond the lake into the blackness. They retraced their footprints in the mud to the entrance and the surface, determined to return.

Materials and tools to build a raft were hauled into the cave, piece by piece. At the lake's edge, the explorers assembled a crude wooden raft to float them across an underground lake of unknown distance and depth. Leaving the shoreline, Howe and Wetsel knew that if they capsized, they would be without light for many hours—possibly days—before rescuers would arrive. They also risked death from exposure if they became stranded for too long in the damp, cool air with their clothes soaked in 42° water.

The raft carried the explorers an eighth of a mile to the far edge of the lake. Beyond the lake the discoveries came in quick succession. They entered into great halls and chambers—the most beautiful in the caverns. They climbed over high underground mountains of breakdown—collapsed slabs of the limestone ceiling—and through narrow, water-filled canyons. On each trip the explorers made, the cave became more familiar and less

An 1889 S.R. Stoddard photo of a tour group in "Washington's Hall."

foreboding. Eventually they explored nearly a mile and a half of underground passageway, all by the dim, flickering light of a small oil lamp.

"Improvements" to the cave began almost immediately, and Howe's own announcements to the press heralded the find as a "rival to the great Mammoth Cave of Kentucky."

Yates, the newspaperman who had explored Ball's Cave, returned to the Schoharie Valley to visit the new cave discovered by Howe. A portion of his report was picked up by the *Supplement to the Courant*, a features-oriented broadside published every other week for subscribers to the Hartford, Connecticut, *Courant*. On Saturday, September 17, 1843, a small headline announced: "A New Natural Phenomenon. Discovery of a vast cavern in Schoharie, N.Y." *The Courant* reported:

> The new cave is not to be identified with the celebrated "Ball's Cave of Schoharie," but is reported as far exceeding it in vastness, besides being more remarkable in its structure. Mr. Yates, a correspondent [to] the New York *Commercial*, in a long letter of nearly two columns, and from which we make a few extracts, minutely describes this last discovered cave. It is situated in a northeasterly direction from the Schoharie mountains, near the "Cave House" kept by Mr. Lester Howe, a very respectable farmer, who is proud of the cave, being for all that is known to the contrary, its discoverer. Mr. Howe entered it for the first-time last May, since which, he has made numerous explorations, generating on one occasion, a distance of five miles, and yet not coming to a termination!
>
> No name has been given as yet to the cave, but the letter writer remarks: "It might not inappropriately be called The Great Tunnel Cave, but the term Gallery, the primary meaning of which is—a long apartment leading to other rooms—is very expressive. Some would perhaps prefer the name Cataract Cave, as the cataract is truly of its distinguished features."

Earliest Records

The publication of *Geology of New York*, also in 1843, gave widespread publicity to Howe's find, although the discovery was so recent as to prohibit a written description. The lengthy guide to the state's geology also published the only known rendering of the cavern's original, or natural, entrance. The printers' woodcut shows a much-cleared opening in the limestone hillside, which has since been destroyed. (Schoharie's John Gebhard Jr. had been appointed as a special assistant to the geological survey, and he and his father made significant contributions to the first volume of *Geology of New York*. It seems likely they would have been among the early visitors to Howe's Cave.)

By the end of the year, Howe and Wetsel had cleared the property near the entrance and cleared mud, clay, and stone from the cave's stream passage to make it more easily traversed. From a report published near the turn of the century, comes an example of Howe's Yankee ingenuity:

> A . . . subterranean river was the agent that made the cavern; but it had afterward obstructed it [the cave entrance] with debris.
>
> Mr. Howe hit on an ingenuous plan for utilizing the water. He first loosened the clay, gravel, and broken rocks; then stopping other outlets he flooded the main channel and thus forced the stream to sweep out its own deposits.

Howe purchased the property from Wetsel in February 1843, reportedly for $100. The land records use the name "Howe's Cataract Cave" in the description of the transaction.

At age thirty-three, Lester Howe opened Howe's Cave as the nation's third show cave. What became of Henry Wetsel is not part of the historical record, and Wetsel is rarely mentioned in any connection with the Howe Caverns of today. Nearby, Wetsel Hollow Road still winds its way from his former property over the hill to the village of Schoharie.

Eight Hours Underground

The earliest paid explorations through Howe's Cave were real adventures. Howe charged 50¢—the equivalent today of about $15.00—to take early guests on a torchlit, eight-to-ten-hour tour, through such fancifully named chambers as the "Washington Hall," "Cataract Hall," "Music Hall," and "Congress Hall," over "Jehosephat's Valley" and up the "Rocky Mountains," through "Fat Man's Misery" and down "the Devil's Gangway."

Often to their chagrin and/or amusement, visitors were provided with clothing suitable for the caverns' trip through mud, clay, and 42° water.

A party of four romantically inclined young men and women from the nearby Catskills toured the cave in the late 1800s and described the attire in the September 7, 1888, issue of *The Dairyman*:

> Everyone carried a lantern and wore a look of determination. It was not considered necessary to add to the attractiveness of the gentlemen. They were provided with straw hats, cowhide shoes, ungainly overalls, and blouses.
>
> The ladies formed a blooming spectacle in navy blue flannel suits, cut after the most approved pattern and trimmed with white braid. Their costume is best described by saying that it was perfectly adapted to the slippery paths and rugged climbing which the cave affords.

Then, as now, the highlight of the tour was an eighth-of-a-mile boat ride across a crystal-clear underground lake. At the end of the lake, a unique column of rock formations—created by both stalactites and stalagmites—produces an eerie, melodious tone when struck. Howe called it "The Harp," and visitors were awed by it and the other wonders of Howe's underground world.

From an historical perspective, it is understandable that after an eight-hour tour the cavern's dimensions would be greatly exaggerated by visitors. Many wrote that the "rival of Mammoth Cave" was at least seven miles in length.

It is also understandable that many thought Howe to be a little bit odd. "An eccentric farmer," an appellation given him by more than one newspaper reporter of the period, appears to have stuck and does so to this day.

Considering the magnitude of his discovery and its worldwide renown, we know relatively little about the life of Lester Howe. Throughout his life he displayed the traditional Yankee traits of foresight, ingenuity, inventiveness, and stubbornness. Howe was tall and thin and was considered handsome, if somewhat somber in appearance. He had piercing eyes and a steely glance, with a small mouth partially hidden by a full goatee and mustache. His ears and forehead were prominent, his nose and cheekbones straight and sharp.

In 1851, educator Simeon North (1802–1884) wrote of Howe: "Out of his cave he was awkward and uneasy, like a sailor on pavements; but no sooner were its rocky walls about him than he straightened into a commanding presence."

In the same account, published in *The Knickerbocker Magazine*, a New York City–based monthly, North added, "We could now help ourselves to a reason why his chin was badly neglected (a reference to Howe's beard), why his eyes glared so strangely in the dismal lamplight, why his back was so partial to a sordid garment. It was so that he might impersonate the Stygian ferryman, so as to fill out the description of Virgil: 'His eyes are flames; A dirty robe hangs from his shoulder.'"

The combination of mystery, danger, and the formidable Howe was undoubtedly awe-inspiring, if not on occasion downright frightening. From an April 1857 account:

> On entering the cave, we had passed the tunnel of stones thinly covered with water; now the stream had risen so high that there was only a foot of space between its surface and the floor of the passage...
>
> Howe drew near, and so held his lamp that we could clearly see the torrent rushing through the tunnel. "There," said Howe, "we must either wade through the passage or retrace our steps and pass

the night within the cave." The water was fast rising, and in twenty minutes would fill the tunnel.

On another occasion in the cave, Howe, "the thunderer," had "petrified his guests into speechlessness." Then, from under his arm, "brought a mysterious box, shaped like a baby's coffin, from which he took out a violin." Howe, the fiddler, made the caverns' visitors "caper about him in wild excitement, his music went to the heels, and the magic of the place transformed the humble instrument into something divine . . . Our spirits [were] buoyed by the music."

The King's Corridor, also from the 1889 photo book by S.R. Stoddard. For more on Stoddard's work, see Section IV, Chapter 2.

The Rival of Mammoth Cave

*Led by Torchlight, Tourists Spend Eight Hours Underground;
"Pip's Journal"*

"In the cave you forget that there is an outer world somewhere above you. The hours have no meaning; time ceases to be; no thought of labor, no sense of responsibility, no twinge of conscience intrudes to suggest the existence you have left. You walk in some limbo beyond the confines of actual life, yet no nearer the world of spirits."

— A visitor to Mammoth Cave, 1855

Despite the rigors of the daylong caverns trip by torchlight, Howe's Cave, in the remote hills of Schoharie County, became an overnight success. By 1845 the Howe family's small wood-frame "Cave House" had an addition built to accommodate the growing number of guests.

A decade later, a writer for *The National Magazine* recorded some of his first impressions:

> The traveler is very glad to see a rude gate having "Howe's Cave" painted in great letters upon one of its bars, and still more glad, when, having turned aside from the main road and crossed a little strip of more smiling landscape, he alights at the door of the hotel, and receives the friendly hospitalities of the great cave explorer.

Travelers would write their impressions of their visit to Howe's Cave in the hotel's guest log, *The Cave Register*. The register has been preserved by the Schoharie County Historical Society.

Students from Union College in Schenectady, described their tour of May 28, 1849: ". . . entered the Cave with Mr. Howe precisely at Midnight . . . Two of the party . . . ascended to the top of the ladder in the Rotunda, and each fired off a Roman candle . . . The party reached the mouth of the cave at 7 o'clock next morning."

Some of the visitors' reactions were written in verse, such as this from October 10, 1852, by an unknown tourist:

I have been in the Cave, well what of that
I've put on the breeches and a dirty old hat
I've groped lamp in hand as far as one can
And came out a wiser and dirtier man.

Warren Howe (a distant relative) writes that "visitors seemed unanimous in their appreciation of Lester Howe for his personally guided tours." The following comments of June 16, 1848, are thought typical:

We had hoped after coming out of the Cave to give some sort of sketch, but it beggars all description. All we can do is return our sincere thanks to Mr. Howe, who conducted us in our journey of five hours to the Rotunda in the very best manner. The reader may rest assured the visit is every way worthy of the time, expense, and trouble.

Lester Howe apparently had a great sense of drama and showmanship. In the caverns' farthest point from the entrance, Howe would fire Roman candles up into the high, circular dome called the Rotunda:

. . . threw up a blazing fire-rocket—up, up it went, far out of sight, with a convulsive and appalling noise as if a comet had struck the earth and hurled it down from its sphere, dashing and rending it to fragments; and when gravity brought it again to our feet it had no appearance of having reached the top, for the stick or tail was whole and the rocket itself unshattered. Truly this funnel must reach to an exceeding height.

— from *Supplement to the Courant*, August 22, 1846

On the same tour, the writer also reported, "Mr. Howe gave several blasts upon a tin horn, the sound of which reverberated along the galleries to the cavern, and echo answering echo through the continuous vaults produced feelings that baffle description. 'Twas [as] if a thousand bulls were suddenly let loose around you, roaring and bellowing, and defying each other to the combat."

Howe also used publicity well. Robert Addis, an avid cave explorer and Howe Caverns guide in the 1960s, has what is believed to be the oldest advertising flier in existence from the heyday of old Howe's Cave. Dated July 1, 1855, it is signed "L. Howe, Proprietor." The text is reprinted here:

Howe's Cave

To the lovers of the wonderful and mysterious in nature, this mammoth Cave, second only to the giant Kentuckian, offers greater inducement than any other place or section of the country that can be found within the wide spreading limits of the American continent. Language is altogether inadequate to portray the thousands of curious objects and singular formations that are constantly presenting themselves to the traveler, as he journeys along through the seven miles of the main thoroughfare of this vast interior of earth.

During the twelve years this cave has been open to the public, many of the most scientific minds of Europe and this country have availed themselves of a personal inspection of the same; have written and lectured upon it, and all agree in pronouncing it one of the greatest natural discoveries to be found in the world.

Every year since the period alluded to, vast sums of money have been expended in clearing and widening the more difficult passages, so that now, ladies can go through the entire length of the cave with as much facility as a gentleman, and in nearly every instance they seem to take greater delight in performing the journey than their companions of the opposite sex.

The cave is now in excellent order, and visitors may depend upon being conducted through by faithful and intelligent guides, persons who have a personal interest in promoting the prosperity of the undertaking and making it a place of resort during the summer months, from every part of the United States.

A rare view of the second Cave House Hotel, built around 1847; it burned in January 1872. Labeled a "scene from the Albany and Susquehanna Railroad," it is difficult to pinpoint where this would have been in relation to the cave entrance, although guests were said to enter the cave through the basement of the hotel. This is a February 2020 addition to Bob Holt's huge photo and postcard collection.

Good and ample accommodations are provided for parties at prices that cannot but be satisfactory.

From Albany to Schoharie the distance is thirty-six miles of excellent plank road, over which passengers are conveyed daily by one of the best stage lines in the country.

Private carriages can be obtained at the latter place, from whence it is but five miles to the Cave, over a pleasant road.

Only a few accounts of the early tours have survived, and it is difficult to reconstruct the entire daylong adventure throughout Howe's Cave in the mid- to late 1800s.

Another photo from the 1889 Stoddard series, taken in "Congress Hall" in the old section of the cave. Note the piping for gas lighting that had been installed leading to the underground lake.

Shortly after the start of the Civil War, during the summer of 1861, a vacationer whom we can identify only as "Pip" toured Howe's Cave. Addis has information that suggests Pip may have been young J. Pierpont Morgan, the millionaire banker and philanthropist. "Pip" was Morgan's childhood nickname, although one of which he was not fond, and it was used only among family. The great banker's estate confirmed Addis's inquiries that Morgan, who would have been twenty-four, traveled to upstate New York at the time of "Pip's Travel Log."

At the time of Pip's visit, Howe, then fifty-three, had been in business for nearly twenty years. After eight hours in the cave, it is understandable that Pip's descriptions of the cavern's proportions are greatly exaggerated. Yet the account stands as the most comprehensive description historians have of Howe's Cave before its commercial development in the late 1920s.

It is evident that Pip was in awe of the caverns' strange beauty, and he writes of the cave with great reverence. He was much less impressed with the caverns' tour guide, whom he and his companions gave the uncomplimentary nickname, "Plug." The ensuing match of wits that takes place in the cave is truly humorous. With little exception, the account is published here in its entirety.

"Pip's Travel Log" is reprinted in part from "*A Brief History of Old Howe's Cave—Or How Lester Won in the End*," by Eric Porteus of Fort Edward, New York. The article was published in the Spring/Summer, 1977, edition of the *Schoharie County Historical Review*.

Pip's Travel Log

My Dear friend:

Perhaps you will be interested in an account of my summer vacation in 1861: If so, I shall be happy to relate to you that part of it which is between Monday morning, August 19th and Tuesday morning, August 27th. I will now try to give you as good an idea as I can of Howe's Cave.

The entrance is at the base of a mountain which is covered with woods; it opens into a valley which is almost entirely sur-

The Rival of Mammoth Cave

rounded by rugged hills, there being but one narrow opening through which a small stream of water runs. Perhaps the seclusion and wildness of the place is one reason why the cave was not discovered earlier. The Hotel is built directly above and over the mouth of the Cave, so we have to descend three pairs of stairs to enter it. The proprietor was not at home, and we had for a guide, in his absence, a young man with a short nose, burnt face, hair closely shingled, pants with pockets spacious enough to hold his hands and arms to his elbows, a hat scrimped in the same measurement, eyes close together, and a wiry little mouth which we saw at once would tell us only as many things as we paid him quarters for. Living so far back in the country he labored under the delusion that everybody else lived in a similar locality, and of consequence was as green and conceited as himself. However, three of us were more than a match for one, blunt and rough as he might be, and we succeeded in obtaining such information as we needed. He showed us into the dressing room where were coats, pants, and hats, which he directed us to put on over our other clothes.

They were made of bed ticking, were short and large around, covered with mud and full of holes, and when once on us and the hats added, each thought the others transformed into most forlorn Irishmen of the stamp in appearance and condition of him who said, "he was out of money, out of credit, out of clothes, and in debt." A mirror hung in the room and by standing before it we found we were all in the same predicament, one looking as bad as another. Our guide, who wasn't very choice at all in his use of his words, regarding neither elegance or propriety of speech, said we "looked like the Devil." And I admit that saying we "looked like distress" would be keeping back the truth. While we were dressing, our guide went for lights and brought us small oil lamps about four inches tall with a single wick in each, and which we thought would certainly go out. But "Plug"—the name we gave our guide, said "he knew," and told us to follow him; we obeyed and went down the stairs, and entered the mouth of the cave.

The first room, "Entrance Hall," is 150 feet long, about 30 wide and high enough for a tall man to walk comfortably most of the way. While passing through this Hall, I began to ask questions; and first when the cave was discovered: "In 1842," he replied. "How did it happen to be discovered?" said I. "Well, sir," he replied, "Mr. Howe was hunting for caves and came across this." Queer kind of things to be hunting for I thought, and glanced back at my companions to let Plug thereafter carry on the conversation to suit himself; and he immediately began on the state of the country, and informed us with the utmost nonchalance that "Washington had been taken," which seemed to please my companions, and New tried to inform the young man that event had taken place two weeks previous, which led to a reply from Plug, which also led to some tangled talk on the state of affairs between New and Plug—New however, before it was done, elicited from Plug the fact that in these parts they got their mails at least once a week—whereupon I said to New, but meaning my words for Plug who I did not highly esteem, that as they were very much behind the times there in regard to items of news and news in general, that, even if the world should come to an end I thought it doubtful if they heard of it until a week or two after it happened, and that, as he didn't succeed in making the impression upon a young man's mind which he wished, he had better drop the subject entirely: our coming suddenly to a new spacious hall at once ended our conversation and changed the subject for us . . .

We were now in Washington Hall, which is 300 feet long, about as wide as Entrance Hall, and a good deal higher, from 40 feet upwards. On entering every room, our guide had some little legend to relate either in regard to its being named, or something which the name might suggest, and these stories he told in a kind of singing voice as if he had told them a hundred times before and knew exactly into what chasm to point or throw his light to make his account as thrilling as possible, and the more thrilling they were, the less he believed them: and our stay in each room I noticed corresponded with the length of his stories. "Old Tunnel" and "Giant's Chapel" were next passed, and in the latter place he had rehearsed some old tale from the Arabian Nights, which either Plug believed had actually happened there, or else, what is more probable, [he] himself didn't know to the contrary.

I may as well state that I was positively afraid to enter the cave. New had visited it before and had now no such fears. Plug noticed my timidity, and in his rehearsals added that in such and such places there was great danger, looking sympathizingly [sic] at me always, and the more he added to his stories with a view to increase my alarm, the more bold I became, and after the first mile, felt no sense of danger at all. I would not blame a person for feeling timid at first, for great chasms at your side and massive rocks above stare you in the face and threaten to destroy you, and you can't help believing that they will keep their word.

Our path all the way was by the side of a little stream of clear, cold water and before we had gone half a mile we could hear its babble and its dash; which seemed to gather volume as it approached and broke at length upon our ears like the voices of many waters. We crossed the stream I should think it a hundred times. Sometimes it was in a chasm before us, then spread out across our path, then bubbling merrily at our sides—its far-off rush the while sounded as if cataracts and waterfalls were plunging into some abyss together. A stream of water underground whose source no one knows! Yet it has a mission, feeding some far off spring perhaps—and always busy at its task with a cheerful happy laugh!

We came next to Harlem Tunnel which is 600 feet long. Frequently on the sides of the rooms were large openings which we did not enter, as most of them had not been much explored. Never in my life have I seen such rocks—sometimes piles of one above another, and seemingly ready to fall at a touch, and now grand boulders of immense size pushing out from the walls and almost 40 feet above our heads, which through the openings between we could gaze upwards but only into the darkness. Perhaps you can get some idea if you will recall how on a hot summer day those terrific thunder clouds roll up from the north and west and sometimes meet; they are dark and threatening, and awful; so those huge rocks that I saw pushing out from either side far above me reminded me of those clouds and seemed as black and as terrific as they.

We then passed through Cataract Hall, which is 300 feet long and remarkable for this: about midway there is a small opening in one side which extends far into the mountains; it is not known how far, for it cannot be explored. But here is something most curious; for you put your ear down and listen there and you are astonished and start back. You listen again; and you hear the sound of a cataract pouring over rocks, and you listen till you are filled with awe. There far off in the mountain, where man never was or ever can be, where the Creator's hand and power alone has been, goes down some grand waterfall, just as in the beginning. He made it to go, and not the sight of which, but only the sound thereof man is permitted to enjoy. I tell you, my friend, man begins to feel his littleness in such a place, for all about him are marks of a Power that must have been Almighty. Yet the Hand that opened these chambers, and balanced those stupendous rocks, and marked out the part for the merry brook, and poured the far-off invisible Niagara, still keeps them in their primal order.

Then we pass the "Pool of Siloam" where the water bubbles up like a little fountain and seemed to sparkle with delight as we held our lamps near to it. Then came "Franklin's Hall" where Plug had to repeat the famous saying which old Benjamin gives in one of his letters as the advice of a mother to him when he was quite young. Plug gave it to us in this form—"Then gentlemen I advise

The Rival of Mammoth Cave

you to remember to stoop as you go, and you will miss many hard thumps"—saying it with all the gravity and dignity of grey-headed Mother himself, so that we couldn't help laughing him in the face.

We came next to the "Flood Hall" where the course of Plug must relate how several persons got so far once when the water rose suddenly and they had hardly time to save themselves by flight, and indeed were waist deep in water before they reached the Entrance, adding, of course, as a subject for our contemplation, what we already knew, that there had been a flood a few days before we came, but he thought there was no danger now.

"Congress Hall" had something of interest besides its legend. But in "Music Hall" which follows it there is so much more of interest that I will omit the former.

And first is the echo, whence the name of the Hall. A small aperture in one side is just wide enough to admit one's head, who makes some audible sound with his voice, and oh what a melody!—as if a thousand harps were struck. It matters not if the sound is ugly, echo calls out to echo, and wave follows wave, till melody is all about you. I could have listened to this music for hours. If my soul was filled with awe in "Cataract Hall," it was here calmed by the sweet voice of music more harmonious than tuneful lips or chord of viola or harp.

Editor's note: At this point, Pip's tour leaves the "historical section" of Howe's Cave and proceeds through the half mile of caverns which is now developed as part of the current commercial tour. Much of the historical section has been destroyed by the quarrying of limestone for cement manufacturing.

Pip's account continues:

The second curiosity of "Music Hall"—and "Musical" is the proper name—is its Lake, of considerable length and from 10 to 30 feet deep. [The underground lake has since been named "The Lake of

Venus."—Ed.] It is one of the most remarkable sights I ever saw. Its waters are clear as crystal and its surface is smooth as glass. Here I saw more of a resemblance to the River of Death than language or fancy ever painted for me. I could hear the dip of muffled oar, and looked upon our guide as a second Charon, taking us over to some unknown shore. We strained our eyes to see across it, but the feeble rays of our lamps could not light up the darkness, which hung its thick veil over the farther side. We stood perfectly quiet in the boat, and our lamps hove their light over ten thousand stalactites that hung like clusters of icicles from the ceiling and sparkled above us like a starry sky. Huge blocks of limestone were piled up on our right hand and on our left. We heard yet the same confused rush of water as before, but here in the stillness it came with a mellower and sadder sound. Darkness now began to close in behind us, and we seemed to be imprisoned there. We passed slowly on with none but solemn thoughts, and in a little while reached the farther shore.

At the end of the lake is what is called "Annexation Rock" or "Rock of the World's History"; it is a limestone formation of tremendous size, 20 feet in diameter and 40 feet high. It is egg-shaped—much the largest in the middle. Our guide said these stones form an eighth of an inch in a century!

Then came the "Museum," where all sorts of formations, some the most beautiful ever beheld, both hanging from the ceiling and side formations, which I can describe better after I finish the room. The Museum is one-half mile long.

Then we begin to ascend the "Alps," and here look out for your feet! We seemed to stand man above man, and each clinging to the loosened rocks. The summit is soon reached,—and they are not of very great height, and take their names more from their steepness and roughness, I suppose,—and then came to descent which is as trying and not at all safer,—but we came down perfectly well, and came into the "Bath Room," where there is a pool of water and sundry other arrangements from which a vivid imagination, like Plug's for instance, could easily believe this to be the Bath Room of the spirits who inhabit there. It certainly has to recommend it, that it is the most retired Bath Room I ever saw.

Next came "Pirate's Cave," but I doubt if any pirate ever saw it; indeed it seemed to me it would require more of pirate courage to venture so far underground as we were than to go on the most hazardous trip a sea-robber ever made.

We came next to the "Rocky Mountains"; and in ascending either these or the Alps and I have forgotten which, we were obliged to go up by ladder a part of the way.

We then passed them safely and came down into "Jehosephat's Valley," one mile in length. Like the one of old it is a deep ravine with the steep sides, and the brook running at the bottom. It is said of the old Valley that every nook and every available spot is crowded with the graves of Moslems and Jews who think it is the greatest honor to be buried there, for each expects that Jehovah or Mohammed will come back to that spot to judge the dead at last.

"Miller's Hall" we passed through next, where we pause a moment to take leave of the brook which has been our companion so long, and also of the Cave proper. There is a dark chasm overhead and a deep ravine below where the Cave and brook lose themselves; rocks have fallen here so that the Cave cannot be explored beyond.

Here we turn a little to the right and enter what is called "Winding Way" which fully realizes the import of the name. Solid rocks on either side, very close together, very high in some places and low in others; we twist, and bend, and compromise, and turn, and turn, and turn, round and round it seems, for a long distance; yet the turns are frequent and the formations on the sides so close to our faces, and the stillness is so great, for the rushing of waters has ceased now that we pass the distance pleasantly and rapidly.

We came out into a small room, and I looked down into a dark hole called "The Devil's Gangway"; it is but little larger than a barrel and 30 or 40 feet long, down into which people must go, head foremost, on their hands and knees, and not expect to turn round or back out. Some fat men have been stuck here and have had to do their best to get either one way or the other.

Editor's Note: A previous day's rainfall fills the small passage at this point and makes it impossible to continue into the final section of the cave. Plug describes the cave for Pip and New, and they sit and rest before starting back. Pip estimates they are now "six or seven miles" from the caverns' entrance. They retrace their steps to leave the cave.

Pip continues his account by describing the formations he has seen:

Let me first mention what is called "Lot's Wife," a formation resembling a human body and of life size. Plug did not know whether it was Lot's first or second wife . . .

In one place was a church—most perfectly represented—with towers and windows and doors. Nearby was an organ with its long, large pipes in the middle, and at the ends, shorter ones, and arranged in such perfect order you should think it was real.

There were also many kinds of vegetables represented: here was a bunch of carrots, perfect as if just pulled from a garden; here a bunch of grass or grain hanging like a sheaf without any band with the heads downward. In another place, was a pair of elephant ears, of life size, and in perfect form. Then there was a pair of doves, loving each other as dearly as people pretend doves do. Again, we saw sparrows and other birds.

There was also a soldier with his armor on, as if some mythic hero had arrayed himself for battle and by some strange overmastering power he was suddenly transfigured and transformed into a pillar of stone—to keep a nightly watch for ages. Not far from here were two women in long flowing robes, facing each other and seemingly engaged in earnest conversation.

But I was most interested in the formation which studded the ceiling above the lake. While the guide was fixing the boat, I ran up the bank to let my lamp shine through them and to examine another formation, which, on the whole was the most curious I saw. My companions were down by the lake, and a growl came up to me from Plug "not to touch that thing," to which I paid no attention whatever, and went on and examined the curiosity to my heart's content. If I had been told that it was a wet skin taken immediately from the tannery and hung there by two or three hooks, so that the folds might hang gracefully, I should have believed it. And now I can think of nothing better unto which to liken it: imagine a hairless buffalo robe thoroughly wet and attached to the ceiling of a room by three hooks about six inches apart, and you have as correct an idea as I am able to give you of this formation. Its thickness was uniform; its folds were graceful; and altogether it was the most singular thing I saw. And now imagine the ceiling of a large room studded with icicles so as to describe every conceivable angle and curve and have them of different lengths and you will get some idea of the number and beauty of the formation which hung over the lake. These limestone formations are not as clear as crystal and are not transparent—they are a dingy white; but because they are constantly wet, the rays of our lamps were reflected by them, causing them to appear most beautiful.

At the place where the Cave and brook lose themselves, I went down into the ravine and drank a good draught of the water which was sweet and pure and cool . . .

Editor's Note: Closing his letter, Pip writes quite eloquently of a unique feeling, one shared by all cave explorers, past and present:

There is also this singular factor to be noticed, that a person seldom feels fatigued in the cave; but on the contrary they are strong and rigorous and ready for any amount of action; there is an exhilaration of spirits; and a suppleness of strength imparted to one such as is no where on the surface of the earth. But, when one comes out into the open air, they are ready to rest, for the exhilarating power is gone and a reaction comes . . .

If you should have the least pleasure in reading what I have written, I shall be more than rewarded for the little trouble it has cost me to record these events, in knowing that I have done anything, however slight, for your gratification.

Yours Truly,

"Pip"

A tour group poses in front of the porch on the Cave House prior to entering the cave, about 1900. Note the tour guide, center, holding his oil lamp.

The Rival of Mammoth Cave

In this 1906 photo, an unidentified explorer sits next to the largest stalagmite in Howe Caverns before the cave's modern development. Today it's known as the "Chinese Pagoda." The 1906 trip was part of a survey of the area's caves conducted by Professor John A. Cook of the New York State Museum. The photo is from a glass-plate negative, courtesy of the state museum.

Lester Howe Loses His Cave

"Lester Howe ... is not a very Smart Man";
Corporate Maneuverings and Quarry Operations Close Howe's Cave

Throughout the 1850s and 1860s, Howe's Cave attracted an increasing number of visitors. The cave became a prominent tourist attraction, known throughout the nation. In 1847, one newspaper reported that "within 20 miles of Albany there is a vast cave, far exceeding in its extent and novelty, the Mammoth Cave of Kentucky."

An advertising flier printed a few years later boasted, "This is one of the most remarkable curiosities in the United States. For extent, beauty, and variety of scenery, it is only equaled by the Mammoth Cave ... with the advantage of being more convenient of access, and without danger."

Just getting to Howe's Cave in the remote Catskills' foothills of Schoharie County presented difficulties that few travelers of today would endure. Schenectady and Albany were, and are, the nearest metropolitan areas. Travelers arrived by boat along the Mohawk or Hudson rivers, or by barge along the relatively new Erie Canal, the primary water routes of the day. The early rail lines also brought travelers to New York's Capital District. (The iron horse would not reach into Schoharie County until 1869.) Then, it was thirty-one miles by coach or horse from Albany along the Great Western Turnpike (now U.S. Route 20) to the northernmost point of Schoharie County. From there dirt ruts and roads led to the Charlotteville Turnpike, the county's main thoroughfare, built of wooden planks laid together in the late 1840s. An uncomfortable carriage ride took travelers the remaining ten or so miles to the caverns' entrance on the hillside. Yet throughout the period, thousands of visitors each year made the trip to Howe's Cave. There are several accounts of there being hundreds of people on a tour at the same time!

Contributing to the caverns' appeal as a preferred vacation spot was the Howe family's Cave House Hotel. While the challenging eight-hour cave tour was only for the adventurous, the pleasant accommodations of the Cave House could be enjoyed by all.

The first Cave House burned to the ground in 1847 from a fire of unrecorded origin. When building its replacement, Howe constructed the northern wing of the hotel directly above the caverns' entrance. (The hotel was more utilitarian than appealing. A recently uncovered photo shows it to have been three stories high and to have had about forty small rooms; it was rectangular in shape without porches or ornate doors and molding.)

Visitors entered the cave through a stairway in the basement of the building, and cool air from the cave circulated up through the lodge. This innovation provided guests of the Cave House, many of them downstate "city folk," with a form of air conditioning—a rare respite from the summer heat.

The air in the cave was the subject of much scientific inquiry and speculation by the learned men of the day. The news correspondent E.F. Yates wrote: "On emerging from the cave, I noticed, as I had done before, on leaving Ball's Cave, the great difference between the air of the cave and of the upper earth." Yates compared the air of the cave to "pure cool water of the living fountain," while the outside air was the "insipid water of the rain-vat."

The nitrous earth of the cave, he theorized, "imparts a healthy action to the respiratory organs, and slightly exhilarates as it invigorates the whole system. As germane to this subject, witness the cures of consumption (tuberculosis) effected by breathing the fumes of nitric acid, and the experiment of Dr. Mitchell, of Kentucky, who being much debilitated and afflicted with a pulmonary complaint, was restored to health by inhaling for a period the air of the celebrated Mammoth Cave of that state."

Sitting on the northern slope of the valley, the Cave House commanded magnificent views of the countryside. Guests could share pleasant conversation or relax in the early summer evenings. The dining rooms were described as spacious and cheerful, and served delicious meals, prepared from the freshest beef, poultry, and dairy products from the neighboring farms. At night, guests were likely entertained by Howe or one of his daughters on the family piano. Lester was remembered in one historical account as an accomplished pianist.

Below the hotel, the caverns provided an inexhaustible supply of cool, crystal water. One older cave country resident wrote much later that a dam was built inside the cave about 2,300 feet from the entrance to create a reservoir to hold an abundant supply. Later owners would boast: "Although we make no special claim as to its medicinal qualities . . . it has been pronounced by competent persons to possess rare properties, having in a number of known cases produced the most beneficial results."

Although a successful man, Howe was regarded as an oddity by his neighbors, and stories of his peculiar behaviors persist to this day. He was referred to in newspaper accounts published as late as the 1930s as an "eccentric genius."

Perhaps the oddest example of Howe's behavior was an 1854 publicity stunt in which he staged his daughter's wedding in the cave. On September 27, Harriet Elgiva Howe wed Hiram Shipman Dewey in a natural loft called the "Bridal Chamber" just within the caverns' entrance to old Howe's Cave. Ironically, Dewey was a surveyor for the coming railroad, the president of which, Joseph Ramsey, would later take control of the Howe's Cave property.

A century later, one of the couple's descendants recalled, "[Harriet] was small and retiring with blue eyes and an abundance of light brown hair. The groom, whose name was tucked away inconspicuously in a corner of the invitation to the wedding reception, was six feet tall, handsome and fun-loving with dark brown hair and deep blue eyes."

It is not as widely known that the Howe's eldest daughter, Huldah Ann, was married in the caverns earlier that year, on August 9, to Henry

Lester Howe Loses His Cave

Northrup, according to Frances Howe Miller, a direct descendant. The Buffalo, New York, *Daily Republic* reported on August 28 that the ceremony took place at 10 PM in the cave, which was "brilliantly illuminated." A display of fireworks in the cave followed, which was "at once pleasing and grand."

The newspaper left one puzzling, unanswered question for historians. The wedding, performed by Reverend Dr. Wells of Schoharie, they wrote, was believed to be the second wedding performed in the cave by the reverend.[1] The first wedding, they reported, "being determined upon the moment," would seem to have been an impromptu underground event.

The first wedding may have been a full two years previously. In Thom Engel's 2014 book, *To Rival Mammoth Cave: Howe's Cave before It Was Howe Caverns*, he writes that the first wedding was on August 18, 1852, and between Peter House of Cherry Valley and Jerusha Catherine Flint of Canajoharie.[1]

Howe's son, Halsey, the youngest of the three Howe children, was married in the cave on October 25, 1865, to Julia C. Redfield. (That Halsey's wedding took place in the cave is a recent addition to the historical records, thanks to the Web site *genealogy trails*. The site references the July 1888 obituary for Lester Howe, published in the *Los Angeles Herald*.)

Considering the influence of the church on family life in the mid-1800s, the weddings in the cave were probably considered near blasphemy. Based on the few stories of Lester Howe's behavior that have been told and retold over more than a century, it seems likely that Lester would not have cared.

The Howe-Dewey wedding became a much-publicized part of the caverns' historic record, perhaps only because a well-preserved sixty-five-year-old invitation to the wedding reception was found in the clay by workers during the 1927–29 development of the cave. A staged photo from 1889 has long passed as the actual event and often been referred to as such by caverns' guides.

Regardless, the cave brides must have been very agile young ladies. The 1889 photo, part of a collection by acclaimed Adirondacks photograph/naturalist S.R. Stoddard (1843–1917), shows the Bridal Chamber was reached by climbing a shaky wooden ladder the height of more than two men. The bride in the Stoddard photo wears white.

Today, the Howe-Dewey wedding is commemorated daily on the cave tour in a "bridal altar" that includes a translucent heart cut from a block of pure calcite found during the caverns' modern improvement.

The couple's first child and the Howes' first granddaughter, Annie Laurie Dewey, was born to Hiram and Harriet in the family's Cave House Hotel on September 12, 1860. Annie was the first of six children the couple would have.

Another story of Howe's eccentricities is that he had his carriage drawn tandem, with two horses in a single file. Howe's reason for doing so, as the story has been handed down, is that he "only wanted to be different"; his neighbors paired their horses side-by-side.

Another Howe tale is that he once advertised an auction to be held at his property; when the grounds were filled with his neighbors and potential bidders, he introduced his eldest daughter, Huldah, who performed on piano.

An elderly resident from the neighboring community of Barnerville told the following story to the former manager of the modern Howe Caverns. As a young boy, he remembered Howe (who must have been quite old) pulling up in his tandem-drawn carriage to the local general store. A minister greeted him, "Good morning Lester. I hear the cock do crow; methinks it will rain." To which Howe replied, "I thought God made the weather, Reverend."

From 1851, there was much talk in the Schoharie Valley of establishing a railroad to connect the Albany rail line to the north to the New York & Erie line to the southwest. The proposed line would run through the Cobleskill, Schenevus and Susquehanna valleys, and the sale of stock would finance a $6-million construction contract. The coming arrival of the iron horse prompted great excitement. Towns along the proposed route, as well as individual subscribers, purchased shares of rail stock in the new Albany & Susquehanna Railroad. It is likely Howe purchased a share or two. They weren't cheap; shares sold for $1,000 each in the 1860s, the equivalent of about $20,000 today.

Lester Howe Loses His Cave

Joseph H. Ramsey, 1815–1894.

After several delays, bitter right-of-way conflicts, and legal entanglements, the Albany & Susquehanna Railroad was completed in 1865. Visitors could arrive and depart from the station established at the hamlet of Howes Cave, an easy two-minute walk or carriage ride to the Cave House Hotel. The number of visitors to the cave increased steadily; Howe's Cave became a leading natural tourist attraction, second only to Niagara Falls. It became perhaps the country's most famous cave, thanks to its proximity to major population centers.

Howe, then in his late fifties, likely prospered. He continued to add to his property holdings and made improvements to the cave. It is believed that he seriously overextended himself in the process.

Cutting through the hillside for the Albany and Susquehanna tracks revealed a high-grade limestone, ideal for manufacturing natural cement. In 1869 a small kiln for processing limestone was established a short distance to the south of the cave entrance; a second firm soon followed (see Section II). Howe added property to this venture and entered into a joint-stock agreement with railroad magnate/politician Joseph H. Ramsey and two other partners, the terms lost to history. Ramsey, as state senator, was instrumental in bringing the railroad to Howes Cave and evidently had been watching the Howes' affairs for some time. He apparently wanted full control of the quarry and cave. An astute businessman, Ramsey, the president of the Albany & Susquehanna, realized there was a huge market for cement, cut stone, and plaster in the building trades. Samples from the Howes Cave road cuts made for the railroad had been sent to state geologists; Ramsey knew a fine grade of natural cement could be made there.

In matters of business, Ramsey was not to be trifled with. In 1869, Ramsey faced a crucial proxy fight led by prominent rail tycoon Jay Gould and "Diamond" Jay Fisk of the Erie Railroad. Financial and legal infighting ensued, and Ramsey emerged the victor.

The business plan of the Howe's Cave Association created for the new enterprise called for substantial capital improvements to the kiln and processing equipment and stepped-up mining efforts in the growing limestone quarry to meet the new demand. In the cave, the free-spending association began an expensive, experimental system of gas lighting from the hotel entrance to the head of the underground lake. A flier of the period boasted Howe's Cave as "The Only Cave in the World Lighted by Gas!" (Some of the pipes, corroded with age, still exist. These can be found in the undeveloped portion of Howe Caverns.)

Lester Howe's beloved cave received less attention, although plans called for the construction of a third, likely more appealing-looking hotel and improvements to the surrounding property.

An early illustration of the third Cave House, from an advertising circular, mid-1870s. This portion of the hotel was made from cut stone. Construction started about 1872.

According to reports, Ramsey on several occasions offered to buy Howe's full interest in the cave outright, but Howe declined. Howe loved the cave and loved entertaining visitors from around the world. Finally, when Howe was fifty-nine and acting increasingly eccentric, Ramsey succeeded.

It is difficult to pinpoint exactly when, and how, Ramsey took control of the Howe's Cave property. Warren Howe writes that it probably took place over a number of years, Lester being the victim of "corporate maneuverings he did not fully understand."

The Ramsey organization saw things differently. In a twenty-four-page promotional description of the cave and hotel, the anonymous author wrote, "He [Howe] sold it at a high figure to the Howe's Cave Association, its present owners, and then retired to a small farm on the opposite side of the valley, where he still lives in peace and quietness."

From the archive collection of today's Howe Caverns, Inc., comes the following advertisement by the Ramsey organization, dated April 25, 1869:

Howe's Cave—Situated on the Albany and Susquehanna Railroad, 38 miles from Albany, N.Y. Trains stop daily. The above-named property having passed out of the hands of its former proprietor into the hands of an association, organized under State Laws, the avenues through the Cave have been much improved, without changing natural position or geological formations, rendering this world-renowned subterranean cavern still more accessible than heretofore. The Cave House, a commodious summer resort for pleasure seekers, has been conveniently arranged, and is ready for the reception of visitors. No pains will be spared to entertain and entirely satisfy the thousands who annually visit the grounds. Experienced guides constantly in attendance to conduct boarders and visitors to the Cave.

Typical of the early industrial period, Ramsey was regarded as a hero of the business age. The event was recorded, condescendingly, in the September 11, 1871, edition of *The Daily Graphic*:

Howe . . . is not a very smart man. That is, he didn't understand joint stock companies. When President Ramsey, of the railroad here, offered Mr. Howe $10,000 for his cave, he refused it; but the next day the good Mr. Ramsey organized a joint stock company, the Howes Cave Company, called the capital stock $100,000, and then paid Mr. Howe $12,000 in stock for his cave. As the cave was all the stock ever paid in, Mr. Ramsey is ahead just $88,000, and Mr. Howe $12,000. This shows power of intellect. It shows how a shrewd honest businessman can always succeed, while dishonest people fail. The poor farmer now hoes his beans and milks his cows on the barren hills, while Mr. Ramsey smokes his "Henry Clay" on the balcony of the Cave House. Such is life.

To put it in perspective, the $10,000 Ramsey first offered Howe would be about $165,000 today. Stock is a paper instrument; its only value is what buyers and sellers agree it to be.

Ramsey had continued successes throughout his career, regardless of his tactics and/or ethics, which were not uncommon at that time. Ramsey was also a founder of The First National Bank of Cobleskill, which a century later became Key Bank, a national banking powerhouse. Management of the Howes Cave Association passed on to his son, Charles, after Joseph Ramsey's death.

To his credit, the elder Ramsey showed some deference to Howe during a July 1872 meeting of the Albany Institute, held at the third Cave House, then just under construction. A member of the all-male club suggested the cave be renamed "Ramsey's Cave," after its "principal proprietor" at that time. The proposal was "heartily applauded," according to the club's minutes. Ramsey shrugged off the suggestion, noting he would not favor ". . . robbing Mr. Howe of the right to perpetuate his own name in connection with the cave he discovered."[2]

Howe retired to his property across the valley, a lush plateau on the mountainside he named the "Garden of Eden." From his front porch, he watched train cars of visitors load and unload at the Howe's Cave depot and watched smoke rise from the cement kilns. It is said he became increasingly bitter. Howe lost control of his cave, but he vowed he would have the last laugh on Mr. Ramsey and the world.

Cobleskill Historian Arthur H. Van Voris (1890–1981) found the following article in a July 1884 edition of *The Cobleskill Herald*. The article was published just four years before Howe's death in 1888 at age seventy-eight. In the article, Howe boasts of having discovered a "bigger and better cave." Perhaps it was an act of retribution, or perhaps it was a benign practical joke born of New England wit. With a single haunting statement, Howe created a mystery that remains unsolved to this day—a lost cavern of unparalleled beauty.

The Garden of Eden property, as described, no longer exists. The Howes' property, home, and farm buildings had long since been abandoned by the time plans were announced for a major interstate highway in the early 1970s. Today, I-88 connects Binghamton and the Susquehanna Valley to the Capital District, with four lanes of highway directly above Howe's Garden of Eden farm.

A measure of Howe's state of mind can be inferred by reading between the lines of the following article. Written fifteen years after the loss of the cave he made famous, it is apparent that Howe, still a little bitter, continued to seek ways to entertain the public and profit from it. In addition to describing the "Garden of Eden Cave," the article also describes new "scientific finds" by Howe and announces plans for a hotel, racetrack, and telephone lines to his Garden of Eden station.

From *The Cobleskill Herald*:

The Garden of Eden was about equidistant from Cobleskill and Schoharie, and was reached by taking a turn in the road to the left, near East Cobleskill . . . One soon reaches a road roofed with overhanging boughs and immense boulders and rocks on the right side, and a beautiful forest on the left. Shortly a clearing of about 20 acres is reached and in the clearing is the famous "Garden of Eden."

On the north side is located Lester Howe's house and barns on the edge of a deep gorge. The cleared space is surrounded by a fine timber belt through which a carriage driveway extends the whole distance of one and one-half miles. The driveway has never been fully completed.

The cleared space is so located that the winds sweep over it without touching, and the surrounding hills and forest protect the fruits and crops from the frost. Hence, semi-tropical crops are quite easily produced here and Mr. Howe has 11 acres of fruit trees, including 300 pear, apple, peaches, plums, quince, plus grapes, raspberries, strawberries, blackberries and currants in profusion.

Sweet potatoes are successfully grown in quantity and early frosts which do damage on nearby properties do no damage here.

The south side of this productive garden is by a fertile hill, where it is always cool and refreshing. The trees are tall and straight and furnish excellent shade. The south side is heavily timbered and fringed along the top with mossy rocks principally of the limestone formation, which erosion has produced a picturesque appearance.

The fine opportunity for study by geologists has been taken advantage of as there are numerous recesses in the rock where, in the hottest weather, the air is comfortably cool, so that one can sit within and enjoy the gorgeous view which extends for miles before the eyes of the viewer. And along the base of these rocks extends a natural and easily traversed path, at its foot the ground is ideal, and here is a delightful picnic spot.

The so-called "Jersey Bed Chamber" is located to the left of the entrance to the grounds. Here Mr. Howe has several superior Jersey cows and calves which retreat into this grove in hot weather, hence its name. This area is a charming gulf through which trickles a silver stream. In the bed and along its side are immense rocks weighing hundreds of tons; all lovers of nature will be well repaid to gaze upon this spot.

Lester Howe Loses His Cave

Mr. Howe has left it just as it was created by Nature, not even removing a single branch from the trees. Velvety moss covers the rocks and logs.

To the west, the grove extends almost to the village or hamlet of East Cobleskill; this part is much higher and is located at the base of the majestic, wooded hill with its splendid view toward Albany and Schenectady.

The air is bracing, and it is here that Lester Howe has surveyed off a one-half mile driving-and-trotting track which will be finished as soon as the crops are harvested. We do not hesitate to state that in no place in the entire state is there a finer view than afforded from this "Garden of Eden."

Among the numerous geological specimens displayed by Mr. Howe is one which he found here, greatly resembling a petrified human head. He found it wedged between the rocks in a deep crevice into which he believes, in ages past, a man fell and could not extricate himself. He thinks it could have been one of the Mound Builders. The nose, mouth, eyes, ears, and jawbone are very plainly defined in this petrified mass. The entire soil seems to rest upon the fossil remains of numerous animals and corals.

Throughout this Garden of Eden, comprising in all some 75 acres, there are many excavations evidently made in ancient time. Into one of these pits, Mr. Howe came upon a stone "vault," which was so recent he had not yet gotten around to fully investigate its nature, and possible contents.

Mr. Howe's connection with the Howe Cave project, which cave he, himself, discovered, did not terminate in any satisfactory and profitable manner for him, so he now has no connection with it, whatsoever.

Here it is that he purchased acreage known today as The Garden of Eden, and he has frequently stated that he has discovered a "bigger and better cave." It is surmised by many that its opening is located somewhere in this territory and it is hoped that one day he will announce to the public and open for inspection his new-found cave, and we wish him and his "Garden of Eden" the best of luck.

Finding a "petrified human head" on Howe's Garden of Eden farm has created a historical mystery, or perhaps speculative tall tale all its own. Howe told the writer it might have "been one of the mound builders," or native Indians of the Schoharie Valley. Could it be, as one local historian suggested, the head of the notorious Seth's Henry, the most brutal of all the Schoharie tribe to fight on the side of the Tories? Seth's Henry's body has never been found. It has been suggested his executioners dumped the body in a "crevice on Terrace Mountain."

The fearsome warrior Seth's Henry returned to the Schoharie Valley following the Revolution. In a 1938 novel set during the war, *Smokefires in Schoharie*, author Don Cameron Shafer wrote that the brutal Seth's Henry was identified when he returned to the valley by his girdle, "woven of soft hair, black hair, brown hair, golden hair, red and silver hair gaily patterned with small, bright beads . . . " and by his war club, which "counted thirty-five notches on one side, that would be the scalps, the forty shallow notches on the other, that must be for the prisoners."

Shafer has his novel's hero, Lt. Vrooman (a common local name), recognize the color of his murdered daughter's hair. Enraged, he suggests to other settlers it is a fine night for "hunting wolves." They follow Seth's Henry out into the night and he, or his remains, are never seen again.

Historian Simms in 1845 gives a different account, perhaps with a grain more truth. He has Schoharie's own real-life Revolutionary War hero, expert marksman Timothy Murphy, do away with Seth's Henry after tracking his whereabouts in the valley for several days. "It was currently believed . . . his bones were left to bleach in the forest." Simms wrote that Seth's Henry was killed with a bullet from the same rifle that sharpshooter Murphy fired to kill British General Fraser at the 1777 Battle of Saratoga. The rifle "ended the career of this crafty chief . . . one of the most blood-thirsty warriors of the Revolution."

Lester Howe Loses His Cave

Both are good yarns, told almost one hundred years apart. Neither mentions a specific location for the remains of Seth's Henry. It should be noted the author isn't entirely clear on whether it is a geologic formation "resembling a human head" or the partial remains of a poor soul who could not pull himself from the crevice on Terrace Mountain.

The remains of Seth's Henry have never been found. Howe's "petrified human head" has also been lost to history.

It is interesting to speculate on the impact Howe's Cave had on the Howe family. Howe certainly felt the exuberance of his 1842 discovery. With great pride he prepared the cave for visitors, not only clearing paths where possible, but also giving names to the cave's great hallways and formations such as "The Music Room," "The Winding Way," and "Annexation Rock." His continued trips into the cave probably caused his wife great concern for his safety, and on occasion his farm chores were probably overlooked while Howe "went cavin."

In the earliest years, the first visitors to the cave were the naturalists, scientists, writers, and other educated men and women of the period. Howe undoubtedly felt ill at ease—"like a sailor on pavement"—in their presence. Born into a farm family in a rural, backwoods area, Howe received no formal education of which there is any record. But Howe's Cave earned for Lester, the "eccentric genius," the respect and admiration of even the most well-educated and well-placed visitor.

Warren Howe writes: "If Lester's loss of ownership [of the cave] bothered him in later years . . . he should not be remembered in this context. Lester's real importance to Howe Caverns was not his discovery and one-time ownership, but his exploration, development, and presentation of that phenomenon to the world. By contrast, what individual or group achieved ultimate ownership is trivial. The latter will pass, but Lester Howe's idea and the efforts he made to make the cave an opportunity for human wonder, delight, and learning will live on."

For more than twenty-five years from the time of its discovery, the world-renowned Howe's Cave was, for the most part, a family business. The Howe family was not made wealthy by the cave, nor left as paupers by Ramsey's shrewd stock manipulations.

Yet there is some satisfaction in knowing that Ramsey's ownership of the commercial portion of the caverns' property eventually turned out to be less than profitable.

As the Ramsey stone quarry began to encroach on the Howes' serene, country farm-like setting, its appeal as a destination may have suffered. A student from the Albany Institute wrote in the July 11, 1872, *Albany Argus*:

> A stone quarry, a massive unfinished stone house high up on the side of the mountain and two or three irregular buildings compose the entire place.
>
> Here refreshments were served, consisting of tea and a coffee of unknown strength, and dressing for the occasion began. Blue overalls, heavy boots, loose coats and jackets and slouch hats for the men. The ladies are provided with rubbers and waterproofs, and for comfort should wear bloomer pantaloons, coats and hats. The guides distributed numerous lamps.

Historian William E. Roscoe's editorial bias throughout his 1882 *History of Schoharie County* is toward the exceptionally well-to-do. Ramsey receives lavish praise, along with a special illustrated biography. He refers to Howe as a "farmer," which is not a bad thing, of course, but he could have as easily referred to him as a respected hotelier or travel host to visitors from around the world. Howe had built and run a successful business for thirty years before Ramsey got control of the property.

Roscoe also sounds condescending when he writes, "Mr. Howe's financial condition was such as to debar him from opening the discovery to the visiting world with that display of advertising etc., which is necessary to an immediate success at the present time . . . " This is not true; visitors were touring Howe's Cave almost immediately after its discovery. And it was not, as Roscoe wrote, "long weary years before its wonders were advertised and the cavern made easy of access by blasting and removing debris, that for ages had been crumbling from the ceiling through the action of frost and water at near the entrance."

Lester Howe Loses His Cave

The construction of the third Cave House hotel began before Howe had fully transferred his property to the Howe's Cave Association. Finishing the work and then adding to this imposing gothic Victorian structure of limestone, the Ramsey organization built an extensive addition of wood that more than tripled the hotel's size. It was renamed the Howe's Cave Pavilion Hotel. Spacious rooms were added for fancy dress balls, billiards, and even indoor bowling; the lawns were manicured for tennis and croquet; and a livery stable provided "good vehicles and horses at reasonable rates."

Historian Roscoe writes that the Pavilion Hotel was completed after additions in the winter of 1880 and 1881.

It is evident the Ramsey organization felt the resort business would be more profitable than the cave business. Lester Howe had charged 50¢ for a tour through his cave; rates for room and board at the Pavilion Hotel were $2.50 per day and $10 to $15 per week. The hotel's promotional literature was reprinted in the *Schoharie County Historical Review* in the Fall–Winter issue of 1971.

Excerpts from the *Review*:

> The Pavilion Hotel, which has been erected with an eye single to the health and comfort of its patrons, fully realizing that in doing this its popularity and success is assured. It is constructed both of stone and wood, is three stories in height, and so arranged, both interior and exterior, that the most exacting person cannot take exception. The sleeping-rooms are all large and elegantly furnished. Many are arranged en suite, with private parlor, bath, etc.
>
> The house is lit throughout with gas, heated by steam when necessary, every room connected with the office by electric bell, hot and cold baths on every floor . . .
>
> Sanitary Arrangements—This most important feature, as it should, has had special attention, both in and outside the hotel . . .
>
> Entertainment—Fully realizing that our guests will require entertainments of various kinds, we have provided for them, among

other things, a billiard room, bowling alley, and a large hall for charades, concerts, and dances . . .

Excursions will be arranged from time to time to Cooperstown (Otsego Lake), Sharon Springs, Richfield Springs, Saratoga, and other desirable places . . .

While under the head of "entertainment," we must not neglect to call your attention to the most interesting and wonderful feature in close proximity to the hotel, Howes Cave, a full description of which will be found further on. Here the student of nature can find rare studies.

A nominal fee is charged to visit the cave, and guides are furnished at reasonable rates. Dressing rooms, with costumes, and other requisites, have been provided for visitors in the hotel, and immediately at the entrance to the cave.

At the height of the more luxurious accommodations at Howe's Cave, about 1880–1900, the Pavilion Hotel more than tripled the size of the stone Cave House, at the left.

Lester Howe Loses His Cave

It is interesting to speculate on the author's reason for including the following in his/her description of the guide services: "No extortion is practiced or allowed in this particular."

From the period 1890 through the turn of the century, the number of visitors to Howe's Cave gradually declined. As cement manufacture went on in high gear, a small community of managers and quarry workers with their families sprung up as the hamlet that simply became Howes Cave (no apostrophe). Immigrant labor was not uncommon; many of the quarry jobs paid low wages for work in hazardous conditions.

As the number of visitors declined, the narrow-gauge rail car that had been built to take visitors from the cave entrance to the underground lake was put to other uses. In a descriptive brochure from the period, the Howe's Cave Association explained:

This is the first of two "barrel vaults" that lead visitors into old Howe's Cave. The structure above it is the Cave House Hotel; the second vault is directly over the cave entrance on the other side of the road. It is only in the last few years this second vault has been uncovered.

This road is utilized for bringing out the remarkable deposit of clay that exists in a portion of the cave, [clay] which will be manufactured into building brick and Portland cement of a superior quality.

From the quarries is taken some of the finest building stone in the state, and the stone from the mines is manufactured into the celebrated "Ramsey's hydraulic cement." These mines and quarries are interesting places to visit and are inspected by many persons.

As a young man, Floyd Guernsey of Schoharie worked the quarries for the Ramsey organization. Many years later, in a July 15, 1935, letter to the new corporate owners of Howe Caverns, Guernsey wrote:

The location [of the Pavilion Hotel] was one of the finest places in Schoharie County, unspoiled by the hand of commercialism. Many times in my younger days, I visited the beautiful grounds above the entrance to the caverns [where] the big hotel and extensive grove overlooked the beautiful and wide valley below.

All this . . . has been blasted away and entirely spoiled by the hands of commercialism in establishing a smoky and dirty cement plant. The caverns itself was entered for its valuables and turned to commercial uses.

Guernsey wrote of a little-known attempt by the Ramseys to market the caverns' riches as building materials:

Along the banks of the underground lake, there is a deposit of a fine grade of clay. On investigations by the Ramseys, they concluded that a fine grade of brick could be made out of this clay; so just outside the caverns they established a brick factory.

A narrow track was laid in the caverns up to the clay bank of the lake and a special rail car was made . . . for hauling out the clay. This business was active for about a year. These bricks proved to be worthless—the clay contained lime, which slacked and cracked the brick.

Not a brick was sold.

As a young man, I worked for the company making these bricks... with another man. [I pushed] the little rail car up the clay bank and loaded the clay. [Guernsey and his companion would then climb aboard and ride the gravity-pulled car back out to the cave's entrance.]

It has always been a deep regret to me that I had a hand in desecrating one of Nature's masterpieces.

The coming demise of Howe's Cave around the turn of the century is evident in the 1896 publication *Celebrated American Caves* by the Reverend Horace Hovey. The book was a first in its field, and Hovey became regarded as the father of cave sciences. *Celebrated American Caves* contained the most accurate description of Howe's Cave at the time, as well as detailed reports on the other great touring caves of the period—Mammoth, Luray Caverns in Virginia, Wyandot (Wyandotte) in Indiana, and others.

Hovey visited the cave with his son in 1880. His chapter on Howe's Cave contained many previously undocumented items and corrected several fallacies. While the report was complimentary overall, Hovey's truthful analysis generally dispelled the claim that the size of Howe's Cave rivaled that of Kentucky's Mammoth Cave.

From Hovey's *Celebrated American Caves*:

A degree of disappointment must be confessed as to the entire dimensions of Howe's Cave. Some enthusiastic letter-writer once said that it was twelve miles long. The report on the geology of New York states that it has "been explored for a distance of seven miles and seems to extend further." A clerical friend assured me that it was at least six miles long. It is recorded that one avenue "never has been explored to its full extent, although a party once spent eighteen hours in it, traveling the whole time, and not reaching the end." Finding that the proprietors themselves discredited those statements, and had no objections to my measuring the cave, I accordingly undertook the task, assisted by my son, with this result: that the total combined length of all avenues open to the public is only one mile and three quarters, and that there may be a mile or more of additional by-ways and tortuous crevices never shown to tourists; hence the owners are warranted in their honest advertisement that the entire length is about three miles.[3]

Earlier in the same article, Hovey noted "modern improvements" that had been undertaken by the Howe's Cave Association under Ramsey's ownership, including gas lighting. However:

> So much digging and blasting have been done between the entrance and the reservoir as to detract from the primitive wildness of the cave, and it too much resembles an unfinished railway tunnel. Gas, also, has been introduced, thus far with a pleasing effect ordinarily, though far less picturesque than torches and not free from danger. This appeared on the occasion of my first visit, which was in a company of a party of 400 excursionists, many of whom caught hold of the pipes overhead to steady themselves along difficult paths. This procedure disturbed the flow of gas. A number of jets were extinguished; and although frequently relighted, they could not be kept burning.
> The next day we examined critically the whole system of lighting up the cave in company with Dr. Lewis, the chemist of the Boston Gas Works, our conclusion being that it is safe enough, if the pipes and jets are not tampered with nor allowed to be eaten through by rust. We recommended the substitution of electric lights, which are now used.

In 1898, with cave tours and the resort business declining, the Howe's Cave Association reorganized as the Helderberg Cement Company. Tours through the cave were no longer promoted, yet a continuous succession of owners quarried limestone from the hillside for building stone and cement, under several well-known brands.

In 1900, fire destroyed the huge wooden "Pavilion" portion of the Cave House, and tours through the cave were discouraged. The former Cave House Hotel became a boardinghouse and was later converted into office space; the quarry soon became the largest employer in rural Schoharie County.

No one has documented exactly when, but sometime between 1910 and 1925, the first explosive charge in the limestone walls of the quarry face blasted into Howe's Cave. In a 1934 account for the Altamont Enterprise, local cave entrepreneur D.C. Robinson pegged that date as "about 1910 . . . the Helderburgh [sic] Cement Company quarry ruined the lower end by blasting too close to it."

Over the years, approximately 300 feet of "old" cave have been destroyed, including Washington's Hall and Cataract Hall. Today's visitors see less than one-half of the original underground passage and enter through a manmade entrance about one mile north of the third Cave House Hotel, a portion of which still stands.

From *The Remarkable Howe Caverns Story*, 1990:

> The quarry no longer operates; the windows of the Cave House have nearly all been broken, and crude boards have been hammered into place to discourage entry of trespassers. The floors are littered with the once-important papers of a former cement company, and the plaster walls are broken and peeling. A crude stairway—overgrown with vine, brush, and twisted limbs—leads not more than 30 feet down to the former entrance of old Howe's Cave. If one stands above the opening on a warm summer's evening, a refreshing current of cool cave air can still be felt.
>
> The original entrance is closed and gated. Through a manmade portal of limestone, less than 100 feet of old cave remain, leading to the "Lecture Room," where Howe would greet guests to

prepare them for the challenges of the cave tour ahead. Much of the chamber has collapsed. The floor of the cement quarry truncates the cave.

For many nineteenth-century visitors these passages were the introduction to the natural wonders of Howe's Cave. This section of the cave was described in 1966 in *The Schoharie County Guide*, a publication compiled by the Boston Grotto, a chapter of the National Speleological Society:

> Here and there may be seen bits of handrail and remnants of gas light and gas tubing from the old commercialization. Beyond this giant hall, once known as the Lecture Room, there are many blocks of limestone rubble, the results of quarrying activities within the cave. High on the right-hand wall is a Gothic arch that leads to the "Wine Chamber" and to the "Bridal Chamber" where Lester Howe's daughters [children—Ed.) were married.

About sixty-five feet inside the entrance to old Howe's Cave is another cave entrance, to Barytes Cave. Uncovered during mining operations in 1904, Barytes Cave extends to the northwest approximately one-half mile and was once mined for ore used in paint products. More on Barytes Cave later.

Walking across the quarry leads to the "back door" of Howe Caverns, an artificial opening in the quarry wall where the old section of the cave continues. The passage consists of about 1,800 feet of the natural cave, left undeveloped. (Coming from the other, tourist end of the cave, this is what extends beyond the lake.)

In the mid-1940s, cave explorer and author Clay Perry described his comical attempts to enter the commercial section of Howe Caverns through the manmade, quarry entrance. From his 1948 book, *Underground Empire, Wonders and Tales of New York Caves*:

> We tried it, one hot summer day, led by the intrepid Roger Johnson, who so charmed the guard who was there to keep foolish persons

out, that he just didn't see us drive up to a point near the yawning entrance in the quarry and then walk in. We clambered up the rubble-strewn passage as far as we dared, mischievously intending to surprise Mr. Clymer and Mr. Hall (the cave's managers) by sneaking in like boys under a circus tent.

We had to retreat to the mine, stumbling over fossilized rocks, picking some up and pocketing them, thus plundering the cement company, which doubtless never missed them.

In the mine, the grinding machinery of the cement mill shook the very rock above our heads until we found ourselves looking cautiously up at the old ceiling, again and again. Nothing fell. We came out after walking for perhaps a mile around the many pillars left to support the ceiling.

In the early 1900s, the wonders of world-famous Howe's Cave and the elegant accommodations of its Pavilion Hotel were no longer desired by curiosity seekers. Natural attractions—not just in upstate New York but throughout the country—were declining in popularity among vacationing Americans, with more modern and urbane leisure activities taking their place. The large cities were reasserting themselves as home of the nation's cultural pursuits, and nature could be found, on display, in numerous museums. Sadly, Ball's Cave on the hills outside of Schoharie was the source for many exhibits across the country on caves. So many of the cave's beautiful calcite formations, stalactites, and stalagmites were taken by early explorers—including the two John Gebhards—that today Ball's Cave is virtually barren of these formations.

Just after the turn of the century, an adventurous visitor would occasionally stop at the office of the Helderberg Cement Company to request permission to enter the once-famous cave. If a former guide could be located, permission might be granted and a nominal fee was usually charged. Boys from the neighboring farms would play pirates and robbers inside the entrance, rarely venturing far beyond daylight.

The once-celebrated discovery of Lester Howe had been ravaged by a series of owners, each more concerned with the forward movement of

American industry than the protection of its resources for future generations. Smoke, soot, and cement dust rose from the kilns and boilers around the Howe's Cave quarry and covered the growing community of factory housing with a dingy grey dust.

In the first few decades of the 1900s, upstate New York's rival of Mammoth Cave was gradually being destroyed.

A Note about the Cave House: Several accounts and advertisements give conflicting dates about the construction of the third, stone, Cave House. The second Cave House, built about 1847 by Howe, was destroyed by fire in January 1872. Construction of the stone hotel began almost immediately after that; it is likely it had been planned since at least 1869 by Ramsey and the Howe's Cave Association and that they had been cutting stone from the quarry for it. An advertising flier of that year uses an artist's illustration of the stone hotel that had yet to be built. Soon after construction was completed for the stone hotel—small by lodging standards—the huge Pavilion Hotel was added. This more luxurious wooden addition was destroyed by yet another fire in February 1900.

---

1. We haven't been able to confirm that it was Rev. Doctor Wells that officiated at the 1852 wedding, which opens the possibility of even more weddings being held in the cave during these early years.

2. *To Rival Mammoth Cave: Howe's Cave before It Was Howe Caverns*, by Thom Engel, copyright 2014, published by Lay of the Land Press, Schoharie.

The minutes of the Saturday, July 6, "field meeting" of the Albany Institute describe quite a day. The learned members toured Howe's Cave and followed with a dinner and business meeting at the Cave House afterward.

The group's chairman, Dr. George T. Stevens, noted it was the third field day held along stations of the recently completed Albany & Susquehanna Railroad, and rail president Ramsey was lauded for his support.

Ramsey described for members plans to expand the Cave House to accommodate 100 visitors and include apartments for more permanent residents; he promoted Ramsey's Hydraulic Lime and Cement Company and introduced Schoharie's John Gebhard Jr., then 70, to talk about the cave, the area's geology, and promote the economic benefits of the county's limestone bedrock—as cement and building material.

Gebhard, long-retired as curator of what is today the State Museum, added another wrinkle to the story of the discovery of Howe's Cave. He told the institute that Howe was led "to discover the cave

Lester Howe Loses His Cave

The Pavilion Hotel was destroyed by fire of an undetermined origin in February 1900. The wooden addition was built of wood and lit by gas.

Another view showing the results of the February 1900 fire that destroyed the Pavilion Hotel.

partly by the fact that Indians and outlaws had been known to disappear in a mysterious manner" in that vicinity. Ramsey's minority shareholder, Lester Howe, was apparently not at that meeting.

3. That figure was further refined by Professor John H. Cook, who visited the cave in 1906 for his New York State Museum publication *Limestone Caverns of Eastern New York*, published that following year. Cook documents that the cave was 4,411 feet in length with another 339 feet through the circuitous "Winding Way."

They Bored a Hole in a Hill[1]

*The Tragic Story of Floyd Collins;
a Bold Plan for Reopening Howe's Cave*

Howe's Cave languished from the turn of the century to about 1925. Its beauty and profitability as a tourist attraction had become secondary to the manufacturing of cement.

Lester Howe died in 1888; his wife Lucinda died the following year at her daughter's home in Jefferson City, Missouri. Both were buried in the Cobleskill Cemetery in undistinguished plots. Ironically, according to Warren Howe, the claims against Lester's estate included a statement from the Howe's Cave Association for $7.15. The statement is signed by association secretary and manager Charles H. Ramsey, the son of Joseph Ramsey who had duped Howe out of his cave property.

Warren Howe writes that Lester Howe died solvent, his assets far exceeding his debts. Lucinda appointed son Halsey as administrator of the estate, and most of his parents' furniture and farm implements were sold. Halsey kept some of the farm machinery and had it shipped by rail to his home in Dunkirk, along Lake Erie.

Daughter Harriet Elgiva, who wed railroad surveyor Hiram Dewey in the cave's Bridal Chamber, had moved with her family to Jefferson City shortly after her father died. They, too, were to make a regrettable business decision. Just before the move, Hiram was approached by an innovative forty-something man from a neighboring community and was asked to invest in his latest idea. According to a family story told by the couple's grandson, Charles Dewey, the young man was "going to stop trains with air."

Hiram was no fool, according to the family story, and he politely declined the man's offer. The man was George Westinghouse Jr., inventor of the automatic air brake—patented in 1872—and founder of the multinational corporation that still bears his name. His home and workshop were in nearby Central Bridge, where George Jr. was born in 1846.

The other Howe daughter, Huldah, had also moved with her family to Jefferson City, which sits on the banks of the Missouri River.

Lester's son Halsey John married in 1865 and moved to Dunkirk, where he practiced dentistry. Although he had a successful practice, Halsey John's life was filled with misfortune. He and his wife Julia had two sons die in infancy. For reasons not in the historical records, his wife became incapacitated and lived much of her life as an invalid in hospitals or nursing homes.

Halsey retired a wealthy man. In 1905 he joined his older sisters and their families in Jefferson City. Then sixty-eight, he lived alternately with his niece, Frances Howe Miller, and nephew, Charles E. Dewey. His wife was moved to a Jefferson City convalescence home or hospital.

In 1911, on or about April 15, Halsey John was taken into custody by Cleveland, Ohio, police after he was found wandering about the downtown streets of Cleveland late at night in a "dazed condition." According to the *St. Louis Dispatch*, he carried with him a "tin box" with more than $200,000 in securities, $700 in diamonds, and $400 cash. The box also contained his bank statement, showing a recent deposit of $21,198 in a Jefferson City bank. Halsey told Cleveland police he thought he was in Pittsburgh, headed home to Dunkirk to transact undefined business. Jefferson City relatives were contacted by wire and came to take him home, which he had left the week previously against family concerns.

The Buffalo, New York, *Courier* newspaper, which served the Dunkirk area, also ran the story, headlining it sensationally—"Fortune Found on Dunkirk Man Thought Insane." He apparently suffered from Alzheimer's; the disease was not widely acknowledged or understood at the time, having only been identified a few years previously.

Two years later, Halsey walked away from his nephew's home again, on a Friday afternoon in late June. A June 30, 1913, *Kansas City Journal* article, "Dr. Howe Disappears," briefly noted a search was underway for the elderly dentist and that it was "not the first time" Halsey had gone missing. The St. Louis and Kansas City police had been notified by wire to assist in the search. "He is at times deranged," the *Journal* noted and stated the fear that because of his age, "the heat may prostrate him as it is believed he is on foot."

The Kansas City paper was a few days late. Howe's body was recovered from the Missouri River, only a few blocks from his nephew's home. The coroner's certificate gives a June 27 date for his death, ascribed to "accidental drowning." The retired dentist's hometown paper, the *Dunkirk Evening Observer,* carried the death notice: "Dr. Halsey J. Howe, who practiced dentistry in Dunkirk for forty years and retired a few years ago, was drowned in the Missouri River at Jefferson City, Missouri."

The family burial plot is in the Cobleskill Cemetery, where Halsey John is interred with his parents. Huldah Ann died in 1907 and is believed buried with her husband's family in Pittsfield, Illinois. Harriet Elgiva died in 1909 and is buried with her husband's family in Jefferson City.

Joseph H. Ramsey died at his home in Howes Cave on May 12, 1894. While the location of the Ramsey residence is not known, it seems likely he would have had quarters in the luxurious Pavilion Hotel. After living in Albany for several years and maintaining a legal practice there, the excessively flattering *Noted living Albanians and state officials* in 1891 stated that he returned home "in the vicinity of a spot where hundreds of pilgrims yearly resort to look upon the silent majesty of nature's works in a 'recess of darkness and wonders.'"

Ramsey's ownership of Howe's Cave is mentioned only sparingly—if at all—in any biography. The *New York Times* death notice, published May 15, 1894, noted, "the first half of his life was dedicated to the legal profession, the second gave to railroad affairs."

After his death, his son Charles was named president of the Howe's Cave Association.

They Bored a Hole in a Hill

In January 1925, a tragedy occurred that would eventually spur interest in the rebirth of Howe's Cave. In the Mammoth Cave area of Kentucky, thirty-seven-year-old Floyd Collins set out to find a cave worthy of commercial exploitation. While exploring a tight passage in Sand Cave, Collins became trapped after he dislodged from the ceiling a small, thirty-six-pound rock that locked his foot into a tight crevice. He was less than sixty feet from the surface. During the next sixteen days, the explorer's plight and rescue attempts captured national attention. Trapped and unable to reach the rock that held his leg, Collins could see, hear, and converse with those leading the rescue attempts. Most Americans suffered vicariously their own worst fears through Collins.

Sensational newspaper coverage fanned a national hysteria; hourly bulletins were eagerly awaited by millions listening to the new medium of radio. Even Congress recessed to hear the latest word on Collins's rescue attempts. But the Collins debacle ended in tragedy; the explorer's lifeless body was pulled from Sand Cave on February 15. Floyd Collins immediately joined the ranks of America's folk heroes as "the world's greatest cave explorer." His story has been told many different times, both as fact and fiction, and in many different media, including the 1951 Kirk Douglas film, *Ace in the Hole* (fiction).

During the next several years many of America's commercial caves were opened. Howe Caverns was one of them, pulled from decline and saved from continuing quarrying. The rebirth and successful development 1926–1929 can in large part be attributed to two remarkable individuals—John Mosner of Syracuse and Walter H. Sagendorf (1872–1940) of Saranac Lake.

It was Mosner who proposed the modern engineering developments that make the cave easily accessible—even comfortable—to the average visitor. Mosner was a steam-fitting and heating engineer and the vice president and general manager of Edward P. Bates Company, of Syracuse. Mosner's foresight is described in *The Story of Howe Caverns*, published originally in 1936 by the caverns' corporation.

John Mosner, the Syracuse engineer who proposed in 1926 that the cave's new owners bore a hole in a hill to regain entrance to the famous attraction.

In his early twenties, Mosner had toured Howe's Cave by torchlight in 1890 and was greatly impressed with its beauty. Mosner also toured the popular caves in Virginia and Kentucky before the turn of the century.

At about the time of the Collins tragedy, electric lighting was being introduced to many of the country's show caves. In 1927, Mosner again toured the commercial caves of the scenic Shenandoah Valley in Virginia, which were now lit by electricity. Mosner's thoughts returned to Howe's Cave.

He remembered how difficult it had been to enter the cave through the nearly half-mile passage from the original entrance to the foot of the lake. And some of that cave had been destroyed or at least made unsafe by work in the quarry. Mosner predicted that Howe's Cave, with a shaft for elevators sunk at the inner end of the cave, and with electric lighting, would become a leading tourist attraction.

Mosner was an innovator, serious in dress and appearance, with a square jaw and large mustache. He talked about the cave incessantly, and plans were formulated and reformulated in his mind many times. He convinced others as well that his plan to revitalize Howe's Cave would work; the caverns corporation's first officers included his boss, D. Cady Fulmer, and Virgil Clymer, a Syracuse attorney. Clymer visited the cave, bringing with him a canoe to paddle across the five-hundred-foot underground lake.

They Bored a Hole in a Hill

Walter Sagendorf provided the organization and business acumen for the Mosner plan. Sagendorf had roots in the Schoharie cave country, but had moved on to become a successful restaurateur as owner of the Berkeley Hotel at Saranac Lake in the Adirondack Mountains. Sagendorf's brother John owned the farm property that extended over much of the cave and on which the caverns' entrance lodge now sits. Both were familiar with old Howe's Cave; as young boys they were among the many neighborhood cowboys, pirates, cops and robbers who played just inside the cave's entrance.

Walter Sagendorf was the organizing force behind Howe Caverns, Inc., created to finance a quarter-million-dollar development with the sale of shares in the closed-stock corporation. Sagendorf was elected first president of the corporation.

John Sagendorf and his wife, Mabel, in this 1929 photo with their four boys. John was the Howe Caverns corporation secretary who owned much of the farmland on which the caverns' estate is situated. The boys, from left, are Allan, Willard, Victor, and Walter.

Throughout 1926 and the first half of 1927, the hamlet of Howes Cave was abuzz with talk of the caverns' reopening. Farmland was being purchased quietly at prevailing prices by persons unknown to the locals. By the time the first news story reached the publisher of the Cobleskill weekly newspaper, more than 1,500 acres to the north of the cave's old entrance had been acquired. Most of the families who sold lands to the caverns' developers remained on their lands. Many of them, their descendants, and their relatives have been employed by the cave at one time or another. The names VanNatten, Nethaway, Crommie, Rickard, Lawyer, and Zeh are familiar to most residents of this rural community. In all, developers paid $48,996 for the land and rights to the property on the surface above the cave.

For many residents of the small hamlet of Howes Cave, the cavern remained a source of considerable pride despite its deterioration and quarry damage during the years 1900–1925. And throughout the county a sense of local pride maintained the cave's grandeur as being among the nation's biggest and the best. It was just that no one could see it.

The first news to reach the public of the plans for the cave's reopening was announced with an eight-column banner headline on July 28, 1927, in *The Cobleskill Times*. The story generated tremendous excitement and contained many of the rumors circulating the area, many of which later proved to be embarrassingly false.

According to the weekly newspaper, the developers' plans (which the Times estimated at costing more than $1 million) called "for a new hotel, golf links, air-port, and a railway station to be located along the Delaware & Hudson tracks, two miles west of the Howes Cave station and village. The proposed site of the hotel and golf course commands an unusual view of the surrounding hills . . . the new company is understood to have in mind a summer and winter business for visitors to the cave."

The article, headlined "Financial Group Plans to Re-Open Howes Cave," reported the "elaborate plans" of a Syracuse syndicate to "revive the one-time popularity of the cave, which is known to be the second largest in the world." Continuing the local conceit, the newspaper claimed that "compared with large natural caves in Virginia and even the Mammoth Cave in Kentucky, the immense cavern discovered by Howe is believed to possess great wonders and furnishes a strong appeal to tourists."

Throughout the late summer months of 1927, Mosner, Clymer, Fulmer, and others from Syracuse met often with the Sagendorfs and other Howes Cave residents who expressed a speculative interest in the reopening of the caverns.

The developers enlisted the services of Delevan Clarke Robinson (1885–1960), a Cobleskill High School teacher who was considered to be the local authority on the area's caves. Twenty years earlier, Robinson had conducted a survey of Howe's Cave on behalf of the state which, according to the local newspaper, had considered making the caverns property a state park.

Robinson led many of the trips through the cave as developers sought to impress its wonders upon other potential investors. He and his wife Adah lived on hilltop farm property only about two miles from the cave's old entrance.

Delevan Robinson—better known as "D.C." or "Dellie"—had a lifelong interest in caves. For him and his younger brother Paul, "it was part of a sport to go down on ropes into the rock holes in the fields and to explore the caves," niece Helena Ackley recalled. Helena, 96, now living in Colorado Springs, said the boys also found sections of what became Howe Caverns and Secret Caverns, "as probably other daring boys in the neighborhood did."

On October 11, 1927, Howe Caverns Incorporated was organized as a closed-stock corporation under charter of the laws of New York State. Officers were elected, directors were appointed, and a business office was established in Cobleskill. A speculative offering was published to raise $150,000 of the $250,000 the corporation estimated it would take to make the caverns visit "about as easy, comfortable and clean as walking through the streets of a village."

On the corporation's prospectus, Robinson is listed as a director and as general manager, a post he never held. Clymer became the caverns' first

Delevan Clarke Robinson lived on farm property in the cave country and was considered a local expert on Howe's Cave. He led many of the tours for potential investors and was listed as the first "general manager" in the corporation's prospectus. He is best remembered as "D.C."

general manager, with responsibility for the day-to-day operations. Walter Sagendorf was elected president by the corporation's directors, and Fulmer and Walter H. Cluett, of Dobbs Ferry, just north of New York City, were elected vice presidents. Cluett's family fortune had been made in the second half of the nineteenth century in the shirt- and collar-making segment of the garment industry. Cluett, Peabody and Company, and its sister firm, the Arrow Shirt Co., were headquartered in Troy, New York. Cluett was probably brought into the corporation by Walter Sagendorf, whose Saranac Lake restaurant and resort attracted many influential businesspeople.

John Sagendorf was named corporation secretary, and Clymer became the corporation's first treasurer. The first board of directors consisted of the officers and Mosner, as well as E.K. Hall of New York City and Richard Ward of Lawrence, Massachusetts. All are now deceased.

Corporate officers began soliciting shares of stock in the proposed Howe Caverns, offering 1,000 Class A and 3,000 Class B shares, both classes at $100 per share. (Class A was the preferred share, guaranteeing an $8 per year dividend.) In 1927, $100 was a great deal of money; investors were offered the subscription for $10 (10 percent) down, the balance due over the next five months.

John Sagendorf's wife Mabel recalled in 1985, "It wasn't easy to ask people to invest in the caverns—something which most people had never seen—and having no way of knowing how successful it would be. There are very few people living today who know all the work, worry, and sleepless nights that were involved."

Within weeks of incorporating, the new officers of Howe Caverns, Inc. were ready to publicly announce their plans for the cave. More than $60,000 had been raised by the developers, most of it solicited through the sale of shares in the Saranac Lake area, where Ausable Chasm was, and is, a popular natural attraction.

Walter Sagendorf and Clymer outlined plans to the business leaders of Cobleskill on the night of October 27 at a well-attended meeting of the Chamber of Commerce. The presentation, reported by the local newspaper, put to rest many of the rumors that had circulated about the development.

"Denying published reports concerning the erection of a hotel, golf links and other features near the entrance to the cave, Mr. Clymer stated that the company did not intend to go into these phases of the development at all," reads a portion of *The Cobleskill Times* article, "a pavilion for the sale of souvenirs and refreshments being the only building to be constructed."

The front-page story offers insight into the business acumen Sagendorf and Clymer brought to the Howe Caverns corporation. Their presentation to the local business community, in which the two men covered all aspects of the development plans, must have been a long one.

A sampling from *The Cobleskill Times* article:

In his talk to the local business men [sic], Attorney Clymer went at length into the illumination phase of the cave's attractiveness, alluding to the Virginia caverns which are electrically lighted and produce gorgeous effects throughout the huge subterranean area.

Colored lights and flood arrangements at proper locations, where the natural formations are most phenomenal, he said, made the Virginia caverns unusually attractive to tourists. Howe's cave, he declared, was favorably compared with the famous Luray Caverns in Virginia by several authorities, which were quoted from published volumes on the subject of caverns and magazine articles.

"Looking back into the old days before the advent of motor cars and provisions for electric lighting, the cave could not be compared [with the Virginia Caves]," said Mr. Clymer."

Clymer told chamber members that the developers anticipated 200,000 visitors annually, and Clymer's estimate has proven remarkably true over time. After acquiring title to most of the caverns' property, Clymer continued, the corporation's officers surveyed the gate attendance at the three popular Virginia caves—Luray, Shenandoah, and Endless. (Caverns' developers borrowed heavily on the Luray Caverns' entrance lodge when designing their own.)

"The visitors to the three principal caverns were counted as nearly as that could be done, and on one day, the names on the register of one cavern were counted. These check-ups were made for one day in April and two days in August about ten days apart and the results showed approximately between four hundred and five hundred people had entered Endless Caverns in one day, about the same number in Shenandoah Caverns, one hundred of which had entered during the evening. The caverns are lighted by electricity and business is carried on evenings as well as day times.

"As the temperature in the caverns remains practically the same winter and summers, the business goes on 365 days of the year. At Luray Caverns, the visitors were about as many as the other two caverns combined, there being about 300 automobiles at the entrance one day and were 800 visitors another day. We were informed that the visitors to the three caves mentioned totaled 370,000 last year.

"In order to get an idea of the probable number of visitors to Howe Caverns if they were open to the public, a check was made of the travel on the Lee Highway, which is the principal north and south highway nearest the three caverns mentioned, and also of the travel to the Cherry Valley and the Albany-Binghamton highways. The check showed that the two latter roads did fifteen times the traffic on the Lee Highway. We also find the density of the population near Howe Caverns is much greater than near the Virginia caverns. Within a radius of three hundred miles of Howe Caverns there is a population of 26,000,000 people.

"*In other words, about one quarter of the population of the United States is within one day's automobile ride of Howe Caverns.*" [Italics added—Ed.]

Clymer and Sagendorf both stressed the value of the caverns' rebirth to the business interests of Cobleskill and the surrounding area. A $100,000 advertising and public relations campaign was planned, they said, to coincide with the cave's grand opening.

Clymer closed the presentation:

"What I have stated will give you some idea of the tremendous financial advantage which will come to Schoharie County because of the spending thousands of people will do in the villages through which they pass and in which they shop for a few hours or overnight. The farm, merchant, hotel, restaurant, garage, gasoline stations, and bank will be benefitted, and each will be benefitted to the extent that the traveling public is served on a fair basis and in a courteous manner. This is no doubt a trite saying, but it cannot be emphasized too often or too much. If the visitors find attractive stores and display windows, good restaurants and hotels, and he also finds courtesy and moderate prices at those and other places he trades, he will remember it and tell his friends, and the result and benefits will be continued for Schoharie County.

"I would like to see the Chamber of Commerce, when the cave is reopened or before, start through the newspapers a 'be courteous to the public' campaign. I suggest this not because the average individual is lacking in courtesy, but because we have got to do a little more than the average in order to make the traveling public speak of us in favorable terms, for otherwise one town is not different from another.

"And last, now that you have heard of this wonderful business which is to be located in Schoharie County, you should immediately begin to boost and keep on boosting for Howe Caverns and for your own locality."

---

1. From Clay Perry's chapter on Howe Caverns in Underground Empire, Ira J. Friedman Publishing, 1946.

Work Begins Underground

The Engineering Story; Workers Trapped by Sudden Flooding; Roger Mallery, Workers' Hero

Early in 1928, work in the caverns began. The engineering firm of Smith, Golder and Homburger, Inc., of Saranac Lake began surveying the cave; Roger H. Mallery (1896–1954) of Howes Cave was hired to "do preliminary excavating necessary for construction of a new entrance to the cavern." It was an exciting year in the cave's history.

Morris Karker (c. 1901–1983) was part of that first work crew. He later enjoyed the distinction of leading the first tour group of visitors through the revitalized Howe Caverns in 1929, according to his own account. In 1979, a tour guides' reunion was organized to coincide with the caverns corporation's fiftieth anniversary celebration. Karker wrote the following, printed in the commemorative booklet:

Howe Caverns as I Recall

Sometime after 1925, rumors began to circulate in the community of Barnerville (where I then lived) that someone was interested in reopening the cavern as a commercial venture. Some farms and lands had been purchased in a quiet manner at a fair or prevailing price. But the first public knowledge that a determined effort was being made was when in the fall of 1927, Palmer Slingerland, a resident of Bramanville with a minor interest in the development,

Roger H. Mallery, from his pilot's license photo, about 1935. Mallery led the construction crews that opened Howe Caverns, Inc. Photograph courtesy of his son, Roger H. Mallery, Jr.

came into Charley Quackenbush's shop for the purpose of selling stock in the newly formed company. Charley was in accord with the venture, but for various reasons, did not think he could or should invest in it. I was there working with Charley and had no money at all, but knowing that soon our seasonal work would end, I suggested that I would be willing to buy two shares of stock for $200, providing that if and when work started, I would be given employment so as to pay for them out of my earnings. Possibly because the sale of stock was not moving very well and the company was anxious to enlist local support, my offer was accepted.

A short time after January 1, 1928, I was advised to meet Mr. Homburger at the Howes Cave railroad depot and work under his direction. Mr. Homburger was a civil engineer and would survey the cavern. Homburger [Henry] and his crew, John and Jim Kelly, did not appear on the designated train or day. I was directed to help Roger Mallery and his crew construct a pole shaft into the cavern ceiling, to regain access to the cavern; a cave-in had completely sealed off the only access, so it was paramount this obstacle be overcome before something else could be done. The cave-in occurred about 30 feet from the vertical east-face of the quarry and about 400 feet from the former entrance. It was a hazardous task, but it was accomplished.

I suppose Mallery and his crew were accorded appropriate recognition for this feat. During the winter and spring of 1928, Mallery and his crew completed the walks, laying walls [building paths up out of the streambed—Ed.], and removing obstacles, to make the passageways reasonably safe and comfortable. My second day of work was meeting Homburger and his crew and all of us being led by Dellie Robinson for a preliminary exploration of a totally unfamiliar world. My apprehension climaxed during an accident which occurred when we reached the stone dam where the cement company had laid pipe for a water supply. There was a fairly deep pool in the streambed below the dam; to get to the top, a 16-foot plank 12" wide with cleats nailed on it [to prevent slipping] had been placed there. Dellie, a large man, over six feet and weighing over 200 pounds, led the way. Midway up the plank, it broke and he fell completely submerged into the deep pool below. The rest of us were willing to run back but Dellie, dedicated to the task at hand, overcame his chagrin and discomfort, and after emptying his boots of water, completed the tour. We spent the remainder of the day at make-work jobs away from the silence and blackness of a totally unfamiliar experience.

The survey was a slow, time-consuming task. Angle stations had to be determined, then a hole drilled into the rock to insert brass rods securely and a prick-point made to serve as exact point of angle. It was important that the survey be accurate, so that the test hole to locate the shaft for the new entrance be exact. The sightings between stations were numerous and often close, and it was difficult to see the line on the plumb-bob. Echoes made communication nearly impossible. Progress was slow. We were a month on the job, surveying just under one mile of cavern passage.

We then had to survey on the surface, to locate a drill hole for the elevator entrance. This did not take long. By applying simple mathematics, only a few shots were needed, along with accurate measurements to determine the exact position. Unfortunately, a math error was made, and the first test hole was drilled nearly 200 feet south of where it should have been. Much to the delay, embarrassment, and cost of those concerned, the error was corrected, and the next test hole came through the roof of the cavern exactly on target.

Editor's note: Karker's account of the first test drilling conflicts with the surveyor's own report, which was published as part of the corporation's Story of Howe Caverns. *Homburger's account appears later in this chapter. A later review of Homburger's notes and an independent survey by caver/former guide Robert Addis found that the surveyor "flipped a bearing," meaning he wrote NW instead of NE.*

Karker continued:

There were a few weeks when Homburger had no work for me, and I was assigned to the Mallery crew. I recall removing a six- or eight-foot pile of clay in front of the elevator stops, excavating the bridal chamber and the steps leading to it, and removing a 15-foot column of rock standing in the Winding Way. It was an unusual form of erosion, but unfortunately had to be removed to permit easier and faster traffic to the end of the Winding Way. It was hard, messy work, but not without interest and some fun. When spring arrived, I helped Homburger survey for the highway leading up to the lodge. I worked under [a superintendent named] Stickles for nearly a year, until it was nearly completed. John Sagendorf wanted me to train to work as a guide. I had some reservations concerning my aptitude and liking the job, but agreed to try, as it would be an easy, clean, and reasonably secure job.

I acted as a guide for only a month. And when in 1932, I sold the shares I held, plus any dividends that should have accrued, my association with Howe Caverns, Inc. ended. I cherish the memory of that long ago association, with so many men, now nearly all gone, engaged in a venture that has added much to the benefit of our meager rural economy and wish for its continued success.

— Morris Karker, 1979

Working in the caverns under the most difficult of conditions was not without danger. The men were continually wet and muddy, laboring in a constant 52° atmosphere. They lost their balance and fell on slippery mud. Boats capsized in the 42° lake. Bats flew about.

In a 1979 postscript to his original memoirs, Karker recalled catching Homburger in a fall from the top of a "fifteen- or twenty-foot ladder" he rigged over a chasm. "I kept his head and shoulders from striking [the] rocks below, but he hurt one knee badly and lost a toenail in the fall."

As the underground drainage system for surface water from melting snow and heavy spring rains, the underground stream in the cave is subject to seasonal flooding. John Homburger, then eighteen, assisted his older brother Hank. He recalled "brutal wet duckings" when rising floodwaters threatened to close the caverns' entrance and trap workers inside. "Many a time . . . we led [the workers] out, each man grasping the belt or collar of the man in front and being pushed in turn by the nervous co-workers behind."

At about 2 pm on a warm, seasonably wet Monday, April 24, five of Mallery's crew members unexpectedly appeared at the caverns' entrance, their work clothes entirely soaked in ice-cold water. Frantic, the men sounded the alarm—sixteen workers were trapped at the farthest point from the entrance by a quickly rising current. Their exit blocked, the men sought refuge in one of the small construction shacks about a mile from the entrance. If the underground stream continued to rise—and outside weather conditions indicated it would—the stranded workers would be trapped for days, or worse, drown.

Gathering a rope, his lantern, and other supplies, the five-foot-eight Mallery entered the cave. The entrance passage, Washington's Hall, the

Giant's Chapel, and Cataract Hall were easily traveled, the underground stream being siphoned off by numerous outlets deeper in the cave. Reaching the appropriately named "Flood Hall," Mallery could see the danger—water was beginning to fill the passage. Just ahead lay the underground lake. Mallery wondered if he had the strength to pilot the small boat moored there against the increasing current. Fortunately, Mallery was a strong man.

By the time he reached the stranded crew of laborers, the shack in which they had sought refuge was almost completely cut off. In the passage between Mallery and his men, floodwaters came to within inches of the ceiling. Mallery tied the rope to the nearest anchor he could find and dived headfirst into the 42° current. Struggling upstream for about twenty feet, he finally reached the anxious workers. Together, they tied the other end of the rope, and one by one they were able to follow the line to safety. They emerged, cold and frightened, to a cheering crowd of their coworkers. Mallery was a hero, and news of his action reached newspapers around the state.

Most of the workers from out of town bunked at Louise Provost's boardinghouse, which operated for a few years in the former Cave House Hotel. "We had evenings to share our misery of bruised shins, wet clothes, holes torn in our boots, and [sharing] the best ways of removing the taffy-like clay that stuck to our equipment, our bodies, and the walkways," remembered the younger Homburger.

Workers in the cave came across several unique finds. At the base of the elevator shaft, the initials of an explorer with the date "1851" were discovered. Homburger, the caverns' surveyor, reported "one room at the end of a tortuous passage contained millions of bones of bats, covering the floor like pine needles in a forest." They had been preserved in layers of clay to the depth of several feet. An invitation to the wedding reception of Harriet Elgiva Howe, September 27, 1854, was found in the clay in a section of the old cave. This is now on display in the caverns' lodge, along with an 1889 photo recreating that event in the old cave's Bridal Chamber.

On February 16, 1928, *The Cobleskill Times* reported on page one the discovery of a new rotunda in the cave. Although the "discovery" was probably a news item placed by the caverns' corporation, the article accurately described construction work in the cave at that time. Headlined "New Rotunda in Howe Cave," the *Times* story reported:

> Discovery of new chambers in Howe Cave [sic] during the period of reopening and reconditioning ... has been announced by Howe Caverns, Inc. Although the "rotunda" is well known to most who have visited the cave in years past, opening of further subterranean passages, especially of large size, is entirely new.
>
> With a force of fifty men at work, the repairing of walks and bridges in the world-famous Howe Cave is well underway to make visits by tourist during the coming summer not only safe but attractive. A well-drilling machine is putting an eight-inch bore hole 100 feet through the ground at a point where the new entrance to the cavern will be constructed.
>
> Later, electric wires will lead into the cave through this opening and an illuminating system will be completed throughout the interior so that tourists will be permitted to view the wonderful natural formations by the use of colored and flood light with attending grandeur such as produced in only a few other caves in the world, the nearest being located in Virginia.

Contractor Busy

After completing arrangements with Contractor Roger Mallery, the Howe Caverns corporation is making rapid progress on the huge task of reconditioning and reopening the cave which has been closed to the public for many years. It is arranged that the work excavating for the new entrance will proceed simultaneously with the electrical contract and reconditioning of paths and bridges, so that with the coming of late spring it is expected that the cavern will be nearly ready to receive visitors.

Additionally, the *Times* reported that caverns officials predicted the work would take another four months. (It took another year and four months). The corporation's tentative opening date, according to the newspaper, was July 1, 1928.

The article concluded: "A visit to the cave reveals the information that along with the rapid progress in getting the cavern reopened, comes considerable interest on the part of people residing in this part of the country as they are actually 'seeing the dirt fly.'"

In addition to the daring cave rescue of the workers caught by high water, contractor Mallery distinguished himself again during the caverns' commercial development. About three weeks after the rescue, *The Cobleskill Times* carried on page one the following social item:

> Recalling the days of Edward [sic] Howe, discoverer of the famous Howe Cave who arranged the marriage ceremony for his daughter in the Bridal Chamber, a similar event took place in the cavern last week.
>
> On Thursday evening in the same Bridal Chamber, Miss Margaret May Provost, daughter of Mrs. Louise Provost of Howes Cave and Roger Mallery, son of Mr. and Mrs. Clarence S. Mallery of Binghamton, were united in marriage by the Rev. F.M Hagadorn of Cobleskill.
>
> Those witnessing the ceremony were the bride's mother, Mrs. Louis Provost, her brother, Francis Provost, and John J. Sagendorf, one of the directors of Howe Caverns, Inc.
>
> Mr. Mallery is the contractor in charge of the work of cleaning up the cave. Mrs. Mallery was graduated from Cobleskill high school in the class of 1922 and from New York State College for Teachers class of 1927.
>
> Following a honeymoon in New York City and other points of interest they will reside in Howes Cave.

There was a lighter side to working in the cave as well. Newcomers were taunted by their coworkers with wild tales of bottomless pits and of

Work Begins Underground

monsters and blind fish that attacked workers from the lake. Snakes, frogs, and worms found their way into lunch boxes. Backpacks often carried an additional brick or two. Rumors were circulated that payday would be pushed back by a week, or that workers would be paid in stock certificates that would be honored at the end of fifty years.

Few visiting the cave today can fully appreciate the magnitude of the commercialization of Howe Caverns. Many fail to consider that it took place during a period of American history in which, in many homes, electricity and telephone service were luxuries.

During the past ninety-plus years, only minor changes have been made to the original modernization work undertaken to develop the cave, i.e., the introduction of elevators, level gravel paths, hand railings, and electric lighting. Bricks—88,000 of them—replaced the gravel paths when, in 1938, the cave flooded to the ceiling. The force of the current was so intense it

With the caverns' lodge still under construction in the background, an employee prepares his descent into the cave riding a crane bucket down the 156-foot-deep shaft dug for two elevators.

washed the gravel completely out of the cave. In 1972 a manmade tunnel connected the end of the Winding Way with the elevators to ease traffic through the cave. And 24 miles of the 1928 wiring system were replaced and updated in 1975.

"The Engineering Story," portions of which follow, was first published in the Howe corporations' *Story of Howe Caverns*, copyright 1936. It was written by surveyor Homburger.

Engineering Story

The problem confronting the caverns' developers was that they owned several miles of natural caverns located somewhere underground about a half mile from the then only entrance, but no one knew in what direction or under what property the cave extended. The original entrance passageway presented so many difficulties that only a few persons were interested in exploring it. How could these beautiful caverns be reached by large numbers of people?

The only way was to follow Mr. Mosner's plan, sink a shaft and install elevators at the extreme inner end of the caverns. The place where such a shaft should be built could be located only by surveying the caverns and using the data to definitively fix the spot on the surface. At that point a well hole, called a bore hole, would have to be drilled in order to check the accuracy of the survey, and to determine the character of the ground through which the shaft would be sunk.

The first trip is as fresh to my mind today as the day I took it, and it will always remain one of the outstanding experiences of my life. After a preliminary trip to determine the lay of the land and the conditions, work was begun about January 1928. I had expected to complete the surveying in two, possibly three weeks, little realizing that we would spend three months in what was probably the strangest conditions we shall ever experience.

With my two assistants and some equipment, including flashlights, I again made my way to the caverns by the original opening.

Work Begins Underground

Certainly, that trip was a revelation to us. The entire day was spent in laying out and marking the approximate locations of traverse stations for the survey, and in becoming acquainted with some of the difficulties provided by Dame Nature in a domain over which she alone had presided for untold thousands of years.

During the day I was compelled to change my impression of the magnitude of the job. Some parts of the cavern seemed impossible to survey. There were canyons and gorges, huge piles of rock and debris to get around; a thousand-foot lake to cross with little or no shoreline; raging torrents of water through places with scarcely room to crawl or squeeze; mud, the thickest and slipperiest I had ever encountered; and over all a Stygian darkness which absolutely killed light. To make it more pleasant, thousands of bats flying frantically around, now in darkness, now in a beam of light, the flirting and chirpings giving one an impression of a tomb disturbed, and with vengeful spirits working near.

A very much impressed, tired, wet and mud-covered trio of surveyors emerged that night to the welcome sunset of a beautiful winter day.

The main location traverse to locate the bore hole [on the surface] consisted of fifty-two stations, [measuring about 4,100 feet of cave]; some several hundred feet long, and others only ten or fifteen feet long.

. . . On account of the roar of the water, at times terrific, coupled with the confines of the caverns, it was impossible to communicate directly by voice for any distance and my experienced assistants were certainly a big help because of their understanding of signals and the results [the work] required. They showed a very keen interest, sometimes amounting to genius, in overcoming the obstacles. It is certainly a trying situation, after an hour's work making a setup, slung by rope to operate, to find an assistant doing something wrong, not knowing it and no way to communicate with him. Truly in such a situation, patience is a virtue.

Due to the nature of the surrounding surface country, with its seams and sinkholes, the brook through the caverns formed the drainage medium for several square miles of countryside. Frequently we would enter in the morning after a freezing night and scarcely dampen out boots, and then, unknown to us the outside temperature would moderate, setting loose thousands of rivulets . . . so that some days at 3 o'clock we could not return without swimming or wading to our shoulders in nice, icy water, and a darkness blacker than black.

This condition at first bothered us somewhat, but we discovered that by waiting until five o'clock or later, when there was freezing temperature outside, that the water would subside enough to let us pass comfortably in our hip boots . . .

On one occasion, it took three days of lying and crawling on our stomachs, using a specially made tripod about ten inches high, to survey a single station. We could scarcely breathe for fear of making an impenetrable fog through which we could not see for five feet. Some passages led to rooms so high that our strongest spotlights could not penetrate the gloom of their roofs. The floor of some rooms was so thick with clinging, sticky clay that it would pull off our boots, and to walk at all meant to hold them on with the hands.

During the period of our surveying, the building of the caverns walks was commenced by two shifts of workmen. By this time temporary electric lights had been installed and work on the final lighting scheme was well started . . .

Our traverse closing line was about four thousand feet in length, and after three weeks drilling with a six-inch well drill at the surface point we had designated, we had the satisfaction of seeing the bore hole punch through the caverns roof within three inches of the spot we had marked. It was to me one of the most satisfying sights I ever had, as it was the culmination of six weeks of extremely trying work and it was evidence that our labors had

Work Begins Underground

been carefully and correctly carried on. A careful record of the drill's progress through the ground was kept daily for ground study and provided most useful information . . .

Practically all of the wire and cable [for the caverns lighting system], many miles of it, were passed down through the bore hole and rewound on a reel on the floor of the caverns. All the cement and most of the gravel used in making the railing ties were poured down the same hole. A telephone line, the only means of communication with the surface, was later threaded through the opening and proved to be of inestimable value. Later when the bore hole was no longer needed to pass material through the caverns, and the elevator shaft was under construction, water lines were put through the bore hole.

The hole is capped today on the surface near the caverns lodge and is a reminder to those who know of the hard work and perseverance necessary to locate that particular point; to us who remember the days of darkness and wet; of the hours of crushing silence; of boats capsizing in the lake; of the wading in the swollen torrents; of the bats flying about and striking our body or face; of the discomforts sustained; but above all the friendships made to last a lifetime.

The contract to dig a 156-foot shaft and install 2 elevators went to John Robertson of Scranton, Pennsylvania. Although others locally had bid for the project, including Mallery's company, Robertson's firm had experience in Pennsylvania's coal country.

Digging the elevator shaft posed a few problems, Karker remembered: "The shaft was dug about twenty feet south of the bore hole. There, an unfortunate rock ledge [was encountered and] dropped off . . . so that instead of finding bedrock close to the surface, as the bore hole indicated, it was not found until 60 to 70 feet down." According to Karker, "thousands of feet of strong and costly timbers" were used to crib the sides of the shaft and keep clay and mud from falling in on the workers.

The elevator shaft, its steel and concrete housing and stairwell (with platforms for emergency stops along the way), were built at a cost of $160,000, or about $1,100 per foot. Thirty-four hundred tons of cement and 105 tons of steel were used to construct the entrance shaft necessary to replace the cave's natural entrance that Lester Howe found on a hillside in 1842.

With the completion of the elevator shaft, work progressed more rapidly; makeshift electric lighting was installed, and a telephone line was strung to let workers talk to those on the surface. Although installation of the two elevators for carrying paying visitors was many months away, a gas-powered winch and hoist carried work crews to and from the surface in a huge bucket. The bucket ride took laborers directly into the main portion of the cave, as we see it today, and saved workers a tremendous amount of time. The shaft eliminated the long, wet, and muddy trip from the old entrance to the work area.

"My husband [John]," said Mabel Sagendorf, "often had to go up and down the shaft in this manner. Once he wanted me to go with him, which I did—a never to be forgotten, interesting ride."

The installation of the elevators marked a major milestone for the developers, and the accomplishment highlighted nearly a year of rugged labor under the most difficult of circumstances. The elevator entrance was the engineering cornerstone of Mosner's plan to revitalize the cave, and its completion signaled for developers the nearing realization of a dream.

A Grand Reopening

Triumph and Tragedy

Contrary to the optimistic newspaper reports, Howe Caverns didn't open that July 1 in 1928. As a result, the fall and winter of that year and the spring of 1929 were busy months for the work crews and staff of the caverns' corporation.

The public's interest in the cave was continually piqued through a series of promotional spots on the new medium of radio, specifically on *The National Home Hour*, broadcast in New York's Capital District on WGY. Businesses along the Susquehanna Trail—the Binghamton-to-Albany highway—were urged by the newspapers to plan and promote for the coming tourist boom. They responded with a vengeance. A local pharmacist advertised, "We prescribe a visit to Howe Caverns." A Bradner's hardware store advertised, "The Sun Never Shines in Howe Caverns . . . but for comfort and appearance, protect your home or business place with awnings."

Arthur Van Voris, a Cobleskill hardware store owner and local historian, assembled a team of explorers for a newspaper series on the "lesser caverns of Schoharie County." In his mid-twenties, Van Voris (1890–1981) was small in stature with a triangular-shaped head that began as a narrow, rounded jaw and chin; a neatly trimmed mustache topped a small mouth, and his head was covered handsomely with thick, curly, auburn hair. He wore wire-rimmed glasses in the fashion of the day.

From the Bright Star Flashlight Company brochure chronicling the 1928 exploration of Nameless Caverns. The Van Voris party believed they were the first to fully explore the cave, but had some help not far from the sinkhole entrance. The caption reads, "Some kindly spirits had left a small sapling-and-slat ladder."

Accompanying Van Voris were Edward A. Rew—an employee of the Cobleskill Post Office—and the three "VanNatten Boys," who owned farm property in the cave country near the Howe Caverns estate. An account of their exploits appeared about once every other week in the Cobleskill newspaper. A promotionally enhanced compilation of the series was published in April 1929 by the Bright Star Flashlight company of Hoboken, New Jersey (Van Voris Hardware carried the company's line of flashlights and batteries). The twelve-page brochure, *Cave Exploring with Bright Star Flashlights*, invited readers to enjoy "the interesting story of these intrepid explorers of the almost unknown caverns in Schoharie County, New York."

The brochure highlights "A Few of the Thousands of Uses of Bright Star Flashlights." Bright Star boasted of an interesting innovation—a belt-and-harness attachment that left "Both Hands Free," according to their advertising materials. A portion of the copy reads, "When the belt is slipped through both loop and ring hangers, light is cast on the ground ahead and both hands are free." (See Section IV, Chapter 4.)

A Grand Reopening

Van Voris and his fellow explorers filed reports on four "lesser caves"—Selleck's, Ball's, Benson's, and Nameless Caverns. In each report, Bright Star flashlights and batteries performed with "eminent perfection . . . always disclosing some new and entrancing detail at each turn of the light."

Van Voris was most impressed with Nameless Caverns, the sinkhole entrance of which lies only about a mile to the northeast of the Howe Caverns entrance lodge. "The glory and grandeur of this cavern!" he wrote. "Continuing for the entire half to three-quarters of a mile to the far end, none of us had ever thought to behold such a gorgeous abundance of stalactites and stalagmites—countless hundreds of them!

"Words are all too inadequate," he concluded, "to describe what we saw in Nameless Cavern." The article on the nameless cave ran as the last of the newspaper series. Van Voris followed it up with an article the following week on Lester Howe's fabled "Garden of Eden" cave.

Meanwhile, at the Howe Caverns estate, the contractors were putting the final touches on the combined entrance lodge, gift shop, and coffee shop. A.J. Brockway of Syracuse was the architect. Before the opening, a promotional brochure described it:

> The entrance building, being erected on the hill which forms the miles of roof of the caverns, commands a 20-mile view of the valley below and the mountains beyond.
>
> The style selected by the architect was inspired by the wonderful old manor houses of England, combining a use of materials to give not only attractiveness, picturesqueness and dignity, but also suggest the welcome so characteristic of Old England. In order that hospitality will be the outstanding impression there is provided a large spacious lounge for the arriving visitors, with high, timbered ceiling such as is found in the Great Hall of the English houses. A huge stone fireplace at one end completes the picture of contentment and warm friendliness that will greet all visitors.
>
> The tea room, as well as the lounge, opens directly upon a broad, flower-bordered terrace which extends an hundred feet

across the south front of the building... During the winter months, probably many winter sports will be arranged for.

A private roadway from the highway to the caverns' estate was being constructed. It was an exciting time as final preparations were underway for the grand opening, rescheduled for Memorial Day weekend, May 27, 1929.

Throughout these busy months, Roger Mallery, the general contractor, workers' hero, corporation stockholder, and bridegroom in the caverns' first wedding since 1854, was noticeably absent.

Brochures, maps, and other promotional literature were prepared. In the days before color photographs, color inks were added on the printers' press to black-and-white photographs, creating "natural color" postcards. General Manager Virgil Clymer appears in the earliest of the postcard series, dressed handsomely in poses underground without showing his face.

Chauncey Rickard of Middleburgh was hired to write the spiel for the cave guides, and he prepared the first staff for the grand opening. He was a big, imposing man, with thick, black eyebrows that darkened deeply-set eyes. Rickard was born on a poor back-roads farm and had only a fifth-grade education. He was largely self-educated and nurtured his talents in the theater and fine arts. He wrote flowery essays on local history and composed and directed many community plays and pageants.

Rickard was a student of mythology, and much of the caverns' original presentation included obscure references to the gods of ancient Greece. Many of the paying customers in the modern cave's early years probably found an embarrassing inconsistency between Rickard's jaunty prose and his imposing farm-grown physical stature. His tour-guide presentation included such loquacious tidbits as the following recitation from the boat ride:

> We now come to the Hall of Adonis, a choice portion of the lagoon, for here Venus has been lavish with her beauty. It is the hall of the beloved youth, Adonis, who when killed in the chase, the goddess sprinkled nectar into his blood from which flowers sprang up. The

ceiling of the abode of Adonis is encrusted with formations typifying the flowers which sprang from the beautiful votive.

There were six guides originally, hand-picked by the caverns' management. Rickard was head guide, Morris Karker was next in line, and the others were Will Kilmer, Omer Youmans, Alton (Jim) VanNatten, and Dave Kniskern. Rickard had the unenviable task of training five young recruits, fresh from the farm, to make the eloquent pitch on behalf of their new employer. It was probably no easy job, and Rickard was insistent that the guides' presentation be *exactly* as he had written it. A former caverns' manager recalled the young guides rehearsing their lines to the trees in the woods by the cave, at Rickard's direction.

"With [Rickard's] sense of dramatics, he could really put it over. Many people were impressed, and said so. It was not so with the rest of us. Lacking Chauncey's talent . . . it was an embarrassment to see the look of disgust on visitors' faces," Morris Karker recalled.

The cave guides received $1.92 per day (just less than $20 today), plus tips. This was nearly twice the prevailing wage for farmhands at the time. Guides had to pay $27 for their military-like uniforms, or they could rent them for 10¢ per day from the corporation. Many of the first guides boarded at the VanNatten farm, just across the valley from the lodge.

Finally, the much-awaited grand reopening of Howe's Cave arrived, May 27, 1927. It marked the culmination of more than two years of organization, planning and hard labor, and an expenditure of nearly one-quarter of a million dollars.

Karker, who was there, remembered the excitement, the VIPs and eager crowds, and an ironic twist of fate he said that occurred on the morning of the big day:

Chauncey met his party of officials and notables, gave his welcoming talk, entered the elevator, and went down into the cavern. I followed with the second party. But a lady in Chauncey's part was taken sick while descending . . . and while he was returning with the party, my group, the second party, was the first to make the trip through the cavern.

 If there is any honor to be gained by the distinction, it belongs to Chauncey, and I would not detract from his record by a simple quirk of fate.

Anyone who might have been there on that day has long since passed, and Karker's story is not verifiable.

The corporation's grand opening on that day was a tremendous success. More than 2,000 visitors toured the cave made famous by Lester Howe nearly a century earlier.

With their work completed, the numerous engineers, contractors, laborers, and corporation staff were justifiably proud. Mother Nature's challenges had been accepted and overcome; her handiwork had been cleaned, scrubbed, lit electrically, and made "more beautiful" by modern presentation. The beauty of Howe's Cave could now be seen by well-dressed men and women in "comfortable shoes," according to the brochures.

The cavern's modernization became almost as strong a sales pitch as the beauty of its formations and passages. In a postscript to the chapter "Engineering Story" from the corporation's *Story of Howe Caverns*, the author notes, "This thrilling and fascinating story will impress the reader with the magnitude of perils braved and obstacles overcome to make the

Identified as "Pulpit Rock" in the 1889 photo book by S.R. Stoddard, the formation at right is today simply known as "the keyboard to the pipe organ," the formation just out of frame at the left.

A Grand Reopening

caverns trip pleasant, comfortable, and safe." This engineering accomplishment and its difficulty in the late 1920s is rarely considered today.

Postcards, brochures, and booklets boasted of underground telephone lines. An entire chapter of *The Story of Howe Caverns* is dedicated to the power and voltage requirements of the cave's lighting system. The chapter also notes, "Two Otis gearless traction elevators which are provided with three speeds, convey visitors to and from the caverns level. If desired the elevators can be dropped the length of the shaft in 30 seconds and returned in the same time."

A promotional map, "All Roads Lead to Howe Caverns," printed two decades after the 1929 opening, noted many of the particulars:

A private roadway costing $20,000.
Howe Caverns Lodge, costing $60,000, equipped like a modern country club for your convenience and comfort.

A 1930 view of the same area as the Stoddard photo on the preceding page, taken soon after Howe Caverns, Inc. was opened. A tour guide chats on an underground telephone connecting to the surface.

A huge concrete elevator shaft descending 156 feet into the earth, containing two powerful electric Otis elevators and 245 steel and concrete steps, costing $160,000.

An auxiliary electric plant exclusively for Howe Caverns. More than 1,500 electric lamps provide illumination, many of them 100 candlepower; 26 miles of cable and wire (much of it enclosed in lead to protect it); 35 transformers of 220 KW capacity, and 100 switches. The wiring and switches are installed on the block system, enabling the guides to emphasize interesting spots and places.

Brick walks throughout; steel and concrete bridges. The walks are mainly built beside and above the bed of the underground stream. Along the walks are iron railings, set in concrete ties buried in the walks to insure rigidity. The walks, bridges and railings cost $80,000.

Editor's note: Many of the walks follow the nineteenth-century paths worn by torch-carrying visitors or improved by the previous owners. In some instances, it appears the 1927–29 work crews may have only needed to level and place bricks over these well-worn trails.

On the Lake are two modern flat-bottom boats, each accommodating 19 passengers and two boatsmen who propel the boats with poles like the gondoliers of Venice. The boats were constructed in sections and assembled at the foot of the Lake.

Telephones, located at frequent convenient places throughout the Caverns, connect with the Lodge.

On busy days from 70 to 80 people are employed to look after the comfort of visitors, including a staff of experienced, courteous and uniformed guides. Six men are required to handle the parking of cars.

For perspective, $1 in 1929 is worth a little more than $15 today. So, the $20,000 roadway to the cave would now cost over $300,000.

A Grand Reopening

Original postcards from the decade of the 1930s also proudly picture a unique byproduct of the modernization of Howe's Cave—underground plant life, which grows as a result of the heat generated from the lighting

Underground boating on the Lake of Venus in about 1930, from Howe Caverns' original postcard series.

system. Today, the plant life is overabundant; the maintenance crew regularly pours a mild acid on the lichens, moss, and ferns that have spread throughout the cave.

John Mosner's vision of modernization and Walter Sagendorf's business plan for Howe's Cave proved successful. Again, the cave became a profitable tourist attraction. Upon the reopening of Howe's Cave, the Howe Caverns Corporation immediately established itself as the proprietor of a leading tourist destination. More than 77,000 people toured the cave during its first year of business. At its business peak in the 1960s and early 1970s, nearly a quarter-million people toured the underground wonder each year.

No one is alive today that may have toured both the "old Howe's Cave" and the new, nor do we have any detailed accounts to compare them. But in August 1954, then-Manager Howard Hall was responding to a letter from Mrs. Charles E. Dewey, the wife of Lester Howe's grandson:

"You'll appreciate that with adequate electric lighting in the caverns there is a lot to be seen that was not evident in the old days of torches and lanterns." Hall went on to explain that a son of former cave owner Joseph Ramsey had toured the modern cave and told him, "I've guided thousands of people through the cavern but never really saw it until today."

He continued, "Instead of the single small skiff we used to cross the lake in the old days we now have a 'quarter mile boat ride' with two boats with the capacity of twenty to twenty-two persons and two more boats reserved . . ."

At about 5 am on a very cold Thursday, April 24, 1930, workers at the North American Cement Company in the Howes Cave quarry exploded a 7 ½-ton compound of quarry gelatin, manufactured by DuPont. The charge toppled 160,000 tons of limestone from the hillside of the quarry that surrounds the abandoned caverns' property and former entrance.

As remote as the possibilities of a rock fall were, officials at the North American plant had met recently with Howe Caverns' general manager

A Grand Reopening

Virgil Clymer. The cement company agreed that blasting would take place in the early morning hours and not during the course of the daily tour schedule at the cave.

On that same morning, more than a mile to the northwest, the caverns' electrician, Owen Wallace, thirty-five, entered the cave shortly after 8:00 to conduct his daily inspection of the lighting system, to replace burned-out lamps, and to prepare for the coming business day. As he left the elevators, Wallace punched the switch to light the Vestibule and walked alone down the short flight of steps into the main cave passage. Being alone in a cave, even a well-lit one, can produce an eerie feeling. One hundred and sixty feet below the surface, the only sound is the gurgle of the underground stream. It speaks to you; the imagination can play strange tricks.

Turning on the lights ahead of him as he walked, Wallace quickly passed the Balancing Rock, the Hands of Tobacco, and the Sentinels at the end of a deep ravine. He replaced a few bulbs; nothing seemed unusual. Standing at the base of the Rocky Mountains, he hit the switch to light Titan's Temple, the caverns' largest chamber, which stood about two hundred feet before him. He coughed, cleared his throat, and continued ahead to the southeast. Wallace walked down a flight of stairs and into the Gallery of Titan's Temple and past the Leaning Tower of Pisa, then watched the passage ahead of him light up as he pushed the switch on the handrail. Again, he coughed. This time his throat burned slightly and he felt tired and dizzy. Wallace coughed again and his breathing became difficult. His lungs hurt. He wheezed for breath and turned around. Retreating, he climbed down over the railing, gasping for air and hoping to reach the caverns' stream for some relief.

Seeping its way underground through the intricate maze of the caverns' passages and fissures was a deadly combination of carbon monoxide and nitrous oxide gases—a strange, unforeseen chemical reaction resulting from the cement company's early morning blast and the unseasonably cold weather, which was drawing the colder air into the 52° cave.

Corporation Secretary John Sagendorf, forty-four, entered the cave shortly after 9:00 that morning, also alone. After all, nothing was suspect;

he was probably curious about the progress Wallace was making. Perhaps the electrician had discovered problems with the lighting system. Sagendorf walked quickly through the first quarter-mile of cave.

Around 9:30, still no word had been heard from either man, and tour guide Alton "Jim" VanNatten, twenty-five, entered the cave.

VanNatten returned to the surface at 9:45, choking and gasping, barely able to speak. He found Sagendorf lying face down and unconscious near a flowstone formation known as the Bell of Moscow. Immediately, two other employees, Adam Kennedy and William Wambo, made the descent, hoping to pull the two victims to the surface. The pair returned within minutes, also suffering the effects of the noxious gases that had entered the cave. The three were rushed to an Albany hospital for treatment; they later recovered.

Meanwhile, General Manager Clymer placed emergency calls to the local fire department and rescue squads and to the better-equipped squads from the nearby cities of Schenectady and Albany. Gas masks for the rescue efforts were flown in from Schenectady by Victor Rickard, manager of the Schenectady Airport.

News of the accident spread rapidly throughout the area, and the newspaper offices of *The Cobleskill Times* were besieged with requests for information. The highway leading to the caverns entrance was soon filled with cars, the *Times* reported, "many of them newspapermen from nearby cities seeking information." After all, just four years earlier, William "Skeets" Miller, the reporter who had covered the tragedy of Floyd Collins at Sand Cave, Kentucky, had won a Pulitzer Prize for journalism. Miller, a slight man, was able to reach Collins for a firsthand account and helped rescuers get the trapped man food and drink.

At 11:30, when Captain William Arndt of the Schenectady Fire Department arrived on the scene with four of his men, Wallace had been in the cave for more than 3 ½ hours. With masks and oxygen tanks, Arndt and his men were the first squad adequately equipped to undertake the hazardous attempt. Shortly after noon, Sagendorf was pulled from the cave. Minutes later, Wallace was carried to the surface.

A Grand Reopening

By then, in addition to the caverns' staff, relatives, and scores of the curious, a twenty-man rescue crew had assembled at the caverns' lodge, including some medical personnel. A rescue squad from the North American Cement plant had arrived to offer their assistance. Rickard, the pilot, returned to Schenectady by plane for additional medical supplies, as efforts to resuscitate Sagendorf and Wallace began.

One newspaper reported: "Volunteers worked unceasingly with inhalators, oxygen tanks and applied the prone pressure method of resuscitation." Efforts continued frantically through the early evening. Finally, at 7:45 pm, nearly twelve hours after Wallace had entered the cave, Dr. H. Judson Lipes of Albany pronounced both men dead at the scene.

The following week John Sagendorf, husband and father of four young boys, was buried on his farm property overlooking the Howe Caverns estate.1 The memorial service was well attended and given prominent, front-page coverage by the Cobleskill newspaper under the headline "Sagendorf Funeral Services Largely Attended at Residence Near Cavern He Held Dear." Fitting remarks were offered by the Reverend F.M. Hagadorn, who noted that Sagendorf, as a cave guide, had helped others marvel at the works of God. Upon his death, the reverend pontificated, "A supreme being had become the Guide, leading to higher and greater wonders of God in the life beyond."

The tragedy was filled with sad ironies. Pilot Victor Rickard was Sagendorf's brother-in-law. Electrician Wallace and Sagendorf were cousins. Both lifelong residents of the area, they had played together when young boys in the old section of the cave.

Newspaper accounts listed "Death by Unknown Causes" as the official verdict for the tragedy. A report from Albany noted:

What killed the two men may never be known.

Officials of the [cement] company insist that there are no fumes from the dynamite which would make men helpless so suddenly that they could not reach safety.

The men . . . showed signs of carbon monoxide poisoning, indicating a gas given off by something having been burned. There was also another gas present, some nitrous oxide, it is believed. No natural gases have ever been found in the caverns.

R.A. Bloomsburg, New York Power and Light Company safety superintendent who directed the futile resuscitation work, regards the presence of noxious gases in doubt. "From appearance, it would seem as if carbon monoxide was present," explained Mr. Bloomsburg, "and there may have been some nitrous oxides, which in combination might have produced another gas. Ordinarily the explosion of dynamite does not give off poisonous gases, but the situation in the cave was unusual. It was a cold day; cold air would be drawn into the crevasses and the warm air in the cavern rises. I think the possibility of any gas pocket in the cave is remote."

Since the publication of *The Remarkable Howe Caverns Story* in 1990, another theory has been put forward—that the fumes were from paint, varnish, and polyurethanes being applied to the nearly all-metal boats being constructed on the Lake of Venus. The solvents in varnish are extremely pungent, and the fumes can cause drowsiness, headaches, skin irritation, and dizziness. At high concentrations, a person may become unconscious, suffer respiratory distress, and may even develop pulmonary edema, in which excess fluid collects in the lungs.

The caverns' air was tested the following day and declared safe. Good crowds on Memorial Day weekend of the following month, according to one newspaper, indicated that the caverns' business had not suffered as a result of the tragedy. The chances of the chemical and atmospheric conditions that produced the deadly gases occurring were one in a million. But the caverns' management would take no chances. A large (approximately 12-foot-by-12-foot) ventilating fan was installed beyond the commercial section of the cave, out of sight of visitors. With a reversible motor, the fan forces air from the cave and through it. On hot days, the fan blows cool cave air into the lodge and gift shop, reminiscent of Lester Howe's early efforts to air-condition his Cave House Hotel.

A Grand Reopening

John Sagendorf's widow, Mabel, never remarried. She raised her four sons on the family farm, working as manager of the caverns' gift shop until well into her eighties. All of the boys—Walter, Alan, Victor (named after her brother, the pilot), and Willard—worked at the cave before moving on to other careers, and many of her grandchildren did as well.

Born in 1899, Mabel lived to be 102 years old, dying peacefully in a Cobleskill nursing home in 2001.

Jim VanNatten, the tour guide who managed to return to the surface to sound the alarm, continued a lifelong relationship with the cave. (He was one of the original shareholders.) Even after he retired as manager of the Howe Caverns motel, and until 1970, VanNatten returned to the cave each Memorial Day to conduct the first tour—a personal tribute to the two men lost.

After several years in a Cobleskill nursing home, Jim died in March 1990.

The caverns' Bell of Moscow was renamed. Since the accident, it has been known as the "Giant Beehive."

And it seemed odd at the end of 1930 that the National Portland Cement Association presented the North American Cement Company with a stone monument, "Safety Follows Wisdom," recognizing a perfect safety record for the year at the Howes Cave plant.

But any good lawyer would point out that the deaths occurred in Howe Caverns, and not in the quarry.

Production continued at the cement quarry through a succession of owners, and production peaked at 2 million barrels per year in the 1950s. While there has never been any danger to the caverns, the author can remember during the early 1970s sitting at night in the picnic area of the cave estate, waiting for the sound of the quarry siren. The shrill whistle announced an impending blast; over a mile away, the explosive blast was barely audible. Then, a minute's wait. The ground would rumble and heave slightly before again settling, all within the space of just a few seconds.

Over the years, the cement produced from Howes Cave limestone has received numerous awards for its quality. But in 1976, the quarry was no longer profitable and the Penn-Dixie Cement Company closed it, throwing about 140 employees out of work. A new owner, Flintkote, continued to run the kiln and processing plant but ceased mining operations. In 1986, Flintkote closed the doors of the plant and the 120-year-old cement industry in the hamlet of Howes Cave was ended.

When this book's predecessor, *The Remarkable Howe Caverns Story*, was first published in 1990 and for the three decades that followed, the quarry stood unused, leaving a gaping hole in the countryside many acres wide and about a hundred feet deep. Fires no longer burned in the kiln; steam, gas, and cement dust no longer fumed from the smokestacks and exhaust towers. The engines and turbines that smashed limestone to powder didn't grind. And the dilapidated cement quarry offices in the former Cave House Hotel were no longer occupied by busy managers, secretaries, and clerks.

What the future holds is described in Section III.

---

1. The newspaper reporter may have had this wrong. John and his wife are buried in the nearby Bramanville Prospect Cemetery, in a Sagendorf family plot.

The Opening of Secret Caverns and Knox Cave

Success Encourages Competition

The reopening of Howe Caverns quickly prompted the development of three other commercial caves in the immediate area. Two have since failed; a third, Secret Caverns, remains an active rival for the tourist dollar.

Just months before the grand reopening of Howe's Cave, contractor Roger Mallery left the corporation. According to Karker, Mallery had expected to be awarded the $125,000 contract for excavation and construction of the elevator shafts. "But management deemed he lacked adequate equipment or experience to handle the job, or possibly other reasons, and awarded the job to a Pennsylvania company whose business was constructing mine shafts," Karker wrote.

Mallery, then thirty-three, would open his own cave, he decided. With landowner Leon Lawton, he developed another Schoharie County cave during the winter of 1928 and 1929. Only about two months after the grand reopening at Howe, *The Cobleskill Times* carried this article:

> NEW CAVERN OPEN TO PUBLIC
> SECRET CAVERNS NEAR HOWE CAVE HAS MANY ATTRACTIONS
> Another cavern has been opened to the public in Schoharie County and promises to become one of the important natural wonders in the state. The name has been designated as Secret

Caverns and is located one mile north of Howe Cave village, near Cobleskill.

The cavern was first opened to the public last Sunday after being in preparation since last winter by Roger H. Mallery and Leon Lawton upon whose farm the cave is situated. It is stated that one hundred visitors viewed the cavern on the opening day.

Location Kept Secret

It seems that years ago Lester Howe, the discoverer of the famous Howe Caverns, was accustomed to remark that he knew of a nearby cavern, but insisted on keeping its location a secret, and from this information comes the name of the present cavern, as it is believed that this is the subterranean wonder which Howe had in mind at that time.

By the use of compressed air drills and a certain amount of blasting, the passageways through the cavern have been enlarged so that one may pass through in comfort and with perfect safety, there being but a few places where it is necessary to stoop down.

The natural formations of Secret Caverns are very unusual and many of them resemble objects such as animals and buildings together with an endless number of stalactites and stalagmites in various shapes and patterns. One unusual feature of Secret Caverns is two large grottos which extend upward a distance of over 50 feet.

Opening Secret Cavern will doubtless appeal to additional thousands of visitors to come to Schoharie County and visit this newly opened natural wonder.

Despite the implication of Mallery's claim, Secret Caverns is unlikely to be the "bigger and better cave" Lester Howe boasted of and called his Garden of Eden Cave. Only about a mile away from Howe Caverns, Secret Caverns is an attractive, complex fissure system, typical of New York caves, with high domes and few formations. A 100-foot waterfall was added to

The Opening of Secret Caverns and Knox Cave

Secret Caverns in the 1940s; ponded water from the surface is fed into the cave by means of an electric switch. The two caves are not connected, yet share a similar underground drainage system; fluorescent dye has been traced from Secret Caverns to the stream that exits beneath Howe Caverns' old entrance in the cement plant quarry.

Secret Caverns, too, has an intriguing history, yet not as renowned as that of Howe. Like many Schoharie County caves, it was less of a discovery than a development. It is likely hunters and landowners passed by the gaping sinkhole entrance for centuries; they just didn't see a reason to "go spelunkin.'"

In the early 1900s, it was Richtmyer's Cave, which according to cave explorer/author Clay Perry, the state geologists explored in 1906 and "gave a very discouraging report about."

"It consists of a medium-sized room and a widened joint in Manlius [limestone] which may be followed for 300 feet," the earliest explorers wrote.

Perry continued: "But some roving spelunkers crept into it, and lo and behold: 'We may call this one Nameless Caverns, since it was so little known at the time of our first expedition and so fraught with difficulties that its wonders and beauties had been seen by [only] a few, and we know of no other name. It is our belief that one of our group, our guide on this trip, was the first to fully explore this cavern.' "

The "roving spelunkers" were the four-man party organized by Cobleskill hardware store owner Arthur Van Voris for his newspaper series and his pamphlet on behalf of the Bright Star Flashlight Company. Van Voris, Edward Rew, and two of the "VanNatten Boys" visited the cave in the late months of 1928. In the manuscript that followed the newspaper series, Van Voris wrote, "We understood that it was about to be prepared for a commercial opening, and that the owner did not want any private explorations to be made."

Van Voris felt the Nameless Cavern was the most fascinating of Schoharie County's lesser caves, and he also described it as being the most extensive. His newspaper report of November 22, 1928, provides a rare description of the cave before its development.

Excerpts from *The Cobleskill Times* story by Van Voris and Rew follow:

The surface entrance is easy—a gradual climb down into a rock hole sufficiently large and with natural cut rock steps for a safe footing.

Shortly one leaves the light of day behind and by means of flashlights, after climbing down over rocks a bit further, we found a short wood ladder conveniently placed for getting down a fifteen-foot ledge.

Then very soon, there is another drop off which is left behind if the visitor is willing to slide down a pole that had been placed obliquely from the top of this ledge to the floor beneath.

Continuing down toward the bottom floor of the cavern, progress from this point, without any exaggeration, becomes difficult.

Two views from the original Secret Caverns' postcard series. At left, tourists look up into the two "Golden Domes." At right, an unnamed underground lake in the cave developed in 1930 and still owned and operated by the Mallery family.

The Opening of Secret Caverns and Knox Cave

Then and now: The Secret Caverns entrance lodge, also from the first postcard series; Today's lodge, a bat's gaping maw is ready to welcome visitors. The cave's stairway entrance is behind the lodge to the left.

For the passage from this spot for a distance of many feet is nothing but a crevice through which the visitor must squeeze his way, a few inches at a time, by means of knees and elbows, bearing distinctly in mind the undesirable feature of slipping down through this crevice which yawns beneath his very feet into the depths below.

A certain presence of mind is essential, and progress must be extremely slow; picturing if you can, the visitors wedging themselves forward and down through a crevice so narrow that it is impossible to turn around—face to one rock wall and back to the other and at one particular spot, these walls coming so close together that one almost feels an added inch of breadth would prevent wedging between them.

Judging from the ladder and pole just mentioned, this part of the cavern and the ensuing two hundred feet or more along the rock floor negotiated by crawling, sliding prone and bending down have been known to numerous previous visitors who have managed the narrow crevices.

But here is something in the way of an obstacle which may have prevented further progress into the regions beyond, for any who have come to this point without guides to suggest the possibility of passing the obstacle.

In fact, one of these earlier visitors mentioned to us that he concluded this was the end of possible progress, for at this point the rock wall shelves down to the floor, leaving an archway not more than two feet high and three feet wide and the floor of this archway, if we may call it thus, is no longer rock—but water.

The water is not deep, but it is extremely cold and looking ahead with the flashlight, one cannot determine its duration or possible depth off under the low arch.

But as most of our party had crawled through on the previous visits, and under the urge of their description of rare and beautiful formations to be seen, we rolled up our sleeves and, bending low, crawled on hand and knees through twenty feet of archway, some of us slipping a little water over the tops of our hip boots, we passed this second hazard of pitch darkness and water.

One could easily spend a half day in the cavern and then perhaps not see it all, for there are two long passages. Time did not permit us to follow both, but with continuing along the one selected, to its far end and an underground lake, and coming back to the entrance again, more than three hours were consumed, thus indicating something of its size and interest.

We all estimated that we must have covered well over a mile in distance and we do know that the general direction is south by east, as shown by our compass.

And our memories do not serve us well enough to attempt any

description of the gorgeous sights disclosed to us, in any proper sequence of observing them, so very much did we see and over such a distance covered.

There are no large amphitheaters like the one in Ball's Cave but there were many high rooms and ceilings so lofty that searchlights did not penetrate the darkness to these heights.

Many side rooms opened up along the main passage with round and perfectly smooth walls and conical roofs aptly termed "silos" by some of the party.

Another spot was called the Rocky Mountains on account of deposits of huge boulders which had to be climbed over, massed up in the passage, twenty or thirty feet high.

Water was encountered along much of the main passage but at no place in any uncomfortable depth.

In the high spots, one could walk along just as we are accustomed to do on the ground level. When the ceiling dropped toward the floor level, we had to stoop down and in a couple of instances, it was so low that there was no other choice than to lie flat amongst stones, sand and wet gravel and slide along by means of toes and elbows, a few inches at a slide.

Now and then, the rock floor presented a most unusual appearance, as some of the party remarked, looking exactly like a manmade concrete cellar bottom, studded with round stones. One might imagine that at its period of creation, these round stones had been lodged in the softer bottom which had then hardened into a natural sub-base holding them as secure as if set in concrete.

In the side-wall of one room was a petrified tree stump, hard as flint and yet a stump of a tree as plainly as could be.

Of the formations which we all like to associate in our minds with caverns, our mere words are so inadequate as to dampen any attempt to tell you about them.

Those who had gone as far as the water archway, missed all of these formations as they had not formed on the entrance side.

Soon after this water hazard had been passed, we began to see them—stalactites of all sizes and shapes hanging from the ceiling and stalagmites formed on the floor against the side walls—and as many of them.

Countless scores and hundreds and to substantiate this statement into an unexaggerated fact, may we remark that they were so abundant that in passing along beneath the low walls, I doubt if there was a single member of the party whose flannel shirt was not pretty well shredded by the time we had gotten back to the entrance after the exploration.

Stalactites and stalagmites, gray, brown, white, rose-hued—from their special limestone upbuilding.

Stalagmites weighing several hundred pounds—stalactites as tiny and fragile as a match.

Solid and everlasting. Slender, graceful, feathery, lacy.

Coral-like. Smooth and round, Horned. Branched. Conical.

Here and there along this passage were small grottoes whose entire walls were formed by stalactites growing downward and meeting stalagmites forming up from the floor.

Now and then, where the roof was low and side walls narrow, these formations were so plentiful, hanging down like pointed horns or rough, coral-like thorns that arms and shoulders were scratched despite the protection of clothing to say nothing of the rents and tears caused by pushing along, as mentioned a while back.

It was estimated from the depth at the entrance and the subsequent turnings and windings that this cavern must lie between one hundred and one hundred fifty feet beneath the surface, although no accurate measurement was taken, nor perhaps could be, without a perpendicular opening from the lowest point.

At the far end of the passage, the roof again shelves down almost to the floor, leaving just enough room to slide along lying flat and wriggling forward foot by foot. After some twenty-five feet of such progress, the passage ends at an underground lake and beyond this, it is safe to believe that no man has ever passed and so like one of the famous caverns of the Shenandoah Valley, aside from our own appellation of "Nameless," one might also call it "Endless."

In conclusion, may we remark as previously stated in another article that it has been our good fortune to have, by chance, visited these lesser caverns of Schoharie County exactly according to their size and interest, each one being more fascinating than its predecessor? And in this cavern, one is impelled to leave it with the emotion that he has gazed upon one of the marvels of nature whose untouched grandeur is such that words alone are utterly inadequate to convey the desired impressions.

Today, Secret Caverns is a stalwart rival and does a modest business during the summer months. Over the years the competition has been occasionally intense and acrimonious. Howe Caverns guides have been accused (by Secret Caverns) of vandalism, and Secret Caverns was accused (by Howe Caverns) of redirecting confused tourists from a "Caverns Museum" at a prime location at the nearest major thoroughfare. The feud reached its peak following an August 25, 1995, fire that destroyed the Secret Caverns entrance lodge. Some thought arson was suspected, but no charges were ever filed; suspicions and accusations were made privately.[1] About the same time, a Howe Caverns billboard at a busy intersection warned, "Howe Caverns Tickets Can Only Be Purchased at the Entrance Lodge 2 miles ahead." (The billboard came down, and relations have gotten much better.)

The admission price to Secret Caverns is reasonable, and it is worth a trip.

Secret Caverns and Howe Caverns have been strange bedfellows at times. The Mallery family owned shares in the Howe Caverns corporation. Dozens of signs to lure tourists away from Howe to Secret Caverns are permanently in place on Mallery property along the road to the Howe Caverns entrance.

Mallery and his crews were largely responsible for the clean, modern look at Howe Caverns. An occasional visitor to Howe Caverns will remark that the cave has been overly commercialized: an elevator ride for an entrance; clean brick pathways; concealed lights; and in places, man-made passages. Secret Caverns, while made comfortable for the public, is more of an adventure, as their current billboards and promotional material promise.

Secret Caverns' advertising literature has always subtly criticized the modern engineering developments at Howe and promoted the more natural appeal of the Mallery family's cave. For many years the billboard for Secret Caverns boasted that the cave's "Natural Entrance Saves you Money!" A brochure noted that "the use of a Natural Entrance makes for such a reasonable admission price you can't afford to miss a trip through Secret Caverns."

An older brochure noted, "Explorations proved Secret Caverns the most extensive in the Northeast, and they have been retained in their natural state to be enjoyed by people every year from all over the country."

Even the history of Secret Caverns, as told by the tour guides, takes a satirical jab at Howe Caverns. The Mallery cave, its guides tell visitors, was originally discovered in 1928 by a pair of curious cows, Lucky and Floyd, the latter presumably named after not-so-lucky Kentucky cave explorer Floyd Collins.

The Secret Caverns twist on the Howe tale of discovery takes a dark turn. From the Secret Caverns website: "The bovine duo took an assumedly fatal trip to the bottom of an 85-foot-deep hole, alerting their owners of the caves' existence."

Another tale occasionally told about Secret Caverns' first years is that Mallery, wanting to keep the cave as natural as possible, sent visitors into the cave by lowering them on rope and sending them on a lengthy crawl through a narrow passage to reach the larger inner chambers.

"Today we would call this spelunking," writes German caver Jochen Duckeck on his Web site of show caves, "and it would be a great success." Not many people took those early tours, however, so—Duckeck continued—Mallery installed a wooden walkway, metal ladders, and electric lights. A few years later he replaced the wooden walkway with concrete

The Opening of Secret Caverns and Knox Cave

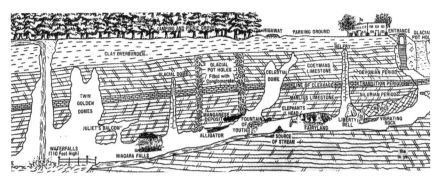

A nice cross-section map of Secret Caverns, from a '60s-era brochure.

and the entrance ladder with 103 steps. That's the Secret Caverns visitors see today.

And this online account isn't the only cave-related publication that describes the caverns' rope entrance and cave crawl. It's also been told periodically in locally produced travel guides—the kind that are mostly advertising.

It seems hard to believe that visitors then might have paid $1.00 for adults and 50¢ for children to be lowered in the cave by rope.

In previous pages it was noted that, prior to the opening, the 1928 Van Voris account described how "The surface entrance is easy—a gradual climb down into a rock hole sufficiently large and with natural cut rock steps for a safe footing... after climbing down over rocks a bit further, we found a short wood ladder conveniently placed for getting down a fifteen-foot ledge.

"Then very soon, there is another drop off which is left behind if the visitor is willing to slide down a pole that had been placed obliquely from the top of this ledge to the floor beneath."

Maybe for safety reasons a rope line was tied around the first tourists to assist their descent. Maybe.

Roger H. Mallery, a Cobleskill attorney, was born the year after his father opened the cave. Mallery, now eighty-nine (in 2020), believed a wooden stairway was originally used in 1929, followed by concrete stairs the next year. He hadn't heard the yarn that visitors were lowered on ropes, either.

What is believed to be the cave's first brochure describes the work to get into the cave's challenging entrance, noting: "... by the use of modern tools, these narrow passages were enlarged ... At the entrance, stairways were built which lead down a winding way to the beautiful cavern below."

A second, more elaborate brochure, with the tagline "Open June 14, 1930" takes pains to note, "In 1929 concrete steps were built leading down into Secret Caverns which were then open to the public."

So, how did the more adventurous tale of rope-climbing and crawling tourists get started?

There's a possible explanation that may have been previously lost to history—people are describing another cave near Secret Caverns!

Cobleskill teacher Delavan ("D.C") Robinson lived on farm property immediately adjacent to Secret Caverns and had become the local expert on caves (much more on D.C. to follow). Years before and up until Howe and Secret opened, Robinson gave tours of wild caves of his own, on family property.

His 96-year-old niece, Helena Ackley of Colorado Springs, wrote in 2020: "There was a long period of time before the Caverns opened, that he [Robinson] organized to take people down our family 'rock hole' on ropes, for pay, and many took him up on it, as spelunkers, and to take photos. It was still going . . . while the Caverns was setting up the ownership lines and barriers to their planned areas."

Robinson would have had to lower his cave visitors sixty-five feet by rope to enter Benson's Cave, one of several caves on the family farm. That cave and some of the surrounding property is now part of the Bensons Cave Preserve of the Northeastern Cave Conservancy.

It is impossible to miss the billboards for Secret Caverns in route to either cave. Folk art lovers may not want to. After a 1986 tornado left many of the original, staid Secret Caverns billboards in splinters, third-generation cave manager R.J. Mallery took a very different approach. The billboards get plenty of attention. One early design featured winged monkeys, evil clowns, skeletons, aliens, and other strange elements, proclaiming, "Things you will not see at Secret Caverns." Currently, there are

The Opening of Secret Caverns and Knox Cave

photos of twenty-seven of the hand-painted billboards featured on the caverns' Web site.

It is also interesting to note the changes made over the years in the ways in which the two caverns are presented to the public. For example, the "Fairyland" formations of Secret Caverns became "The City Hit by Atomic Bomb" after World War II. The anti-war, anti-nuclear power movement of the 1970s probably prompted the change to "City of the Future." In Howe Caverns, the former "Bell of Moscow" became "The Giant Beehive" after the tragedy that took the lives of two employees. A natural pillar on the underground lake in Howe Caverns was removed to add two more boats for busy days, and a manmade tunnel now connects the end of the Winding Way to the Vestibule, again to improve traffic flow on busy days.

The elder Mallery was also the developer of the short-lived Schoharie Caverns, located midway between Schoharie and Gallupville. Believed to have opened in September 1935, Schoharie Caverns didn't last long as a tourist attraction—if it ever really became one.

Welcome to Secret Caverns.

According to the recently unearthed papers of Schoharie Attorney James L. Gage (1908–1991), the energetic Mallery developed the cave on farm property owned by Delos Treadlemire, promising him $20 a month plus a share of any profits from admissions. "He worked all summer with two men . . . cleared away the dirt and rock . . . so that the entrance was large enough to walk into standing up. He put in walks and electric lines, with power generated by a Ford motor set on the hillside. He built a quarter-mile S-shaped road [from Route 443 up to the cave entrance], being careful not to disturb the surrounding farmland."

Gage also wrote (in about 1958) that Mallery constructed a dam across the lower end of the cave, built a boat, and was to have a boat ride for customers immediately after they entered the cave. Perhaps this was to compete with Howe Caverns.

Unfortunately, a torrential flood in late September of that year devastated the area, washing out roads and bridges, flooding the cave and, with a loud boom, collapsed the hillside just outside the entrance.

"Apparently Mallery never returned to continue his work," wrote Gage, "and even left his tools and a wheelbarrow, as well as the wiring, bulbs, Ford motor and generator." Landowner Treadlemire, according to Gage, never received any money from the venture, "but was happy to keep the wheelbarrow."

A brochure was prepared and several small ads appeared over a three-week period in late September–early October 1935. According to the Schoharie Caverns' advertising circular Mallery prepared:

> Here, unlike the other caverns in Schoharie County, the entrance is located on the side of a mountain.
>
> The trip starts with a boat ride on an Underground Lake, then on walks through a beautiful winding passageway well decorated with stalactites and stalagmites together with great masses of flowstone such as are found in the other caverns in the north. At the end of the trip, the cavern is blocked by great formations of flowstone hanging down to a deep pool of water.

The Opening of Secret Caverns and Knox Cave

Electric lights and board walks, together with the absence of stairs, permit old as well as young to explore this cavern almost without effort. The trip requires about 30 minutes.

This cavern has been developed to meet the present-day demand for a cavern with a low admission fee and at the same time the public is given the opportunity at seeing a cavern which is second to none in the North.

Cave explorer/author Clay Perry had a slightly different version in the mid-1940s:

This cave was leased, and efforts made to enlarge the entrance and some of its passages . . . After building plank walks and extending electric lines in waterproof cables far inside, the project failed, due to the depression of the 1930s, and the whole thing was abandoned. The walks and cables, light sockets, and a large engine set up on concrete outside, were left to rust and rot.

Since then, Schoharie Caverns has reverted to its "wild" state. Because its one-half-mile length is mostly walking passage, the cave is popular among novice cave explorers. Some of the electrical wiring and rotted boardwalks remained in the cave up until the late 1970s, when the author first visited. The entrance gate, a wrought-iron spider on its web, remains. (This often confuses Schoharie Caverns with a "Spider Cave," which is outside of Gallupville. Spider Cave has spiders in it, lots of them. Local lore has the spiders getting bigger the farther into the cave one goes.)

It is difficult to imagine a boat ride in any portion of Schoharie Caverns, unless the boats were very narrow. In many places the cave walls close to a width of less than 4 feet. For most of the 2,000-foot caverns' walking passage, the width is between 6 and 8 feet. The water, most of the year, is just ankle deep.

But Mallery set the admission fee at only 35¢ and offered special rates for groups of three or more. Children under twelve accompanied by a parent were admitted without charge, according to the advertising literature.

Attorney Gage was a lifelong friend of the caving community. He also owned the Schoharie Caverns property for several years and considered plans to open it as an attraction himself. After his death, the cave and the surrounding property were donated by Mary and Jennifer Gage in 1994 to the National Speleological Society (see also Chapter 3 in Section IV).

Delevan "D.C." Robinson had been involved with the Howe Caverns project as a surveyor and cave guide for potential investors prior to its reopening. He was listed as general manager in the caverns' corporation's prospectus, but for unknown reasons never held the job.

Robinson had apparently wanted to be a cave owner and promoter since boyhood. His family's farm property—neighboring Secret Caverns—included caves, fissures, sinkholes, and surface depressions, common to the cave country.

His niece, Helena Ackley, 96, said, "The kids played in a rock hole that had a tree trunk fallen over it." With thick and "very heavy barn ropes," the enterprising Delevan—then about fifteen—rigged a hook, counterweight, and pulley to lower others into the caves. He charged a few cents for the homemade adventure, as described earlier.

"These first rope-supported explorations were [kept] secret from the parents," Ackley wrote in a 2020 e-mail. "Mother—Delevan's sister—remembered getting bribes not to tell [on her older brother Dellie]. She would hang out there, above ground, with her dolls and playthings," acting as a lookout.

So, it was no surprise that in 1933—four years after the reopening of Howe's Cave—Robinson purchased and opened Knox Cave, in Albany County just outside of Altamont. He offered Cobleskill's intrepid explorers Arthur Van Voris and Edward Rew the opportunity to join the venture for

From an early Knox Cave brochure, with an artist's rendering of the combined ticket booth and roller-skating rink. The sinkhole entrance to the cave would be behind the lodge, to the left.

$100 each. Robinson sought to raise $1,000 to "do the necessary excavating and build the wooden stairs."

Robinson leased the Knox Cave property from the Truax family in 1933. The large sinkhole entrance was cleared of debris, staircases and electric lighting were installed, and passages were cleared.

"There were inevitable setbacks," wrote Art and Dave Palmer in the September 2004 issue of *The Northeastern Caver*. "The construction supervisor ran off with some funds, and the lighting system was blitzed during a thunderstorm, but tours began on May 30, 1933, and the cave soon became a popular attraction.

"Robinson hired three staff to help him operate the cave and supervise tour guides," the Palmers continued. "In 1935–37 they built a large roller-skating rink next to the entrance. The combination of rink and cave was irresistible in those low-tech days, and they are said to have attracted as many as 1,000 people each weekend. The Robinsons bought the cave and surrounding property in 1937 and took up residence in the old brown house that came with it."

"It [the cave] never was very much developed," wrote Robinson's niece in a letter to family in 2004. "In 1933 and '34 . . . I was nine or ten and my younger brother Carlton [Smith] was seven or eight. We were asked to go into small openings, using a flashlight to see what we could report. I was scared to do this and found it a very tight passage, so much so that I literally had to back out."

But that was then. Knox Cave turned out to be an extensive system. Robinson spent two decades trying to prove that it was the largest cave in the Northeast, according to the 1976 guide, *Caves of Albany County, N.Y.* In the mid- to late 1950s, several discoveries more than doubled the cave's known length at the time, to more than 1,000 feet. Until the roller-skating rink burned and Knox Cave closed in 1958, Robinson had a standing agreement with cavers to waive the 50¢ admission fee, and he offered a $100 reward to any group making additional discoveries.

Knox Cave, named simply for its location in the Town of Knox, also has an interesting history, albeit only partly known. Like Secret Caverns, the cave entrance is a large sinkhole, one that would be hard to miss by Native Americans hundreds of years before the colonists' settlement at Albany. In an early brochure, Robinson notes that the cave was likely "inhabited by a mysterious race" preceding the local Iroquois tribe.

In 1934, explorers probing the undeveloped reaches of the cave found a 2-foot-by-3-foot stone tablet covered with hieratic writings, believed to have been left in the cave by that "mysterious race" around 400 AD.

That Knox Cave was later used by the Iroquois, said Robinson, "is proved by formations deposited over their smoke stains, bones, artifacts and other marks" found in deep recesses.

The Opening of Secret Caverns and Knox Cave

D.C. Robinson shines a light for guests into one of the high domes in his Knox Cave, Albany County, in this rare 1937 photo. Photograph courtesy of Gordon Smith/Hidden River Cave, National Cave Museum and Library, Horse Cave, Kentucky.

In yet another noteworthy discovery, the July 1, 1935, edition of the *Albany Evening News* carried this:

VISITORS TO KNOX CAVE FIND HUGE TOOTH ON ROCKY FLOOR. Visitors gathered for the semi-weekly "squares" at the Knox Cave dancing pavilion near here last Saturday night decided to enter the cave before the orchestra tuned up for the evening and by so doing made a new discovery.

It is in the form of a huge tooth in excellent condition, which has been submitted to geologists and paleontologists at the State Department of Education for a verdict as to its original owner.

The tooth found near the spring, remote from the entrance to the cave, was evidently washed clear of the shale debris from overhead. The management of the cave is anxiously awaiting word from the Albany scientists to which it has been submitted.

Curiosity-seekers could find something to like in any of these three caves during those early years. Howe Caverns, though, was clearly the top draw, and by a long shot. Like the Mallerys at Secret Caverns, Robinson took a few shots at the big competition as well, and along the same battle lines: "My cave is more natural than yours." Perhaps Robinson, like Mallery, was miffed at corporate decisions that affected him. He had hoped to be general manager of Howe Caverns, after all.

The Opening of Secret Caverns and Knox Cave

In Knox Cave, proclaimed an early brochure, "There has been no shifting of formations to increase or improve display and no building of rooms or artificial formations in order to try to reproduce things mentioned in old accounts."

No one of that era would have been more familiar with old Howe's Cave than Robinson, but his subtle jab here was likely too obscure for most who gave any thought about which cave to visit. He is likely referring, in part, to Howe Caverns' "Chinese Pagoda," a huge and beautiful stalagmite that was propped up from where it fell, and the "Leaning Tower of Pisa" and an unnamed companion formation that have both twice been moved to the walkway. The "room and artificial formation" he is likely referring to is the Bridal Altar, with its carved, translucent white calcite heart embedded in the floor. This modern homage to Lester Howe's Bridal Chamber is a rectangular-shaped alcove that would have been unreachable unless steps—about 20 up and 30 down—had been laid up the high embankment of mud, silt, clay, and breakdown.

Photos from D.C. Robinson's Knox Cave in neighboring Albany County. Previous page shows the huge entrance room; above is the stairway leading into the cave entrance. Photos are from the original postcard series.

"Knox Cave has been kept most nearly as nature formed. Paths make easy walking and electric lights show its natural beauty," Robinson wrote for the "Look into the Earth" brochure.

Robinson loved his cave and loved showing it, but it is likely the roller-skating rink and square dances are what paid the bills. "He would drop everything to take even one family through," said his niece.

"If people came into the roller rink to ask about the cave, buy tickets, or arrange to go through the cave, Dellie would have done it right then," Ackley said. His wife Ada would have stayed behind to manage the property while Robinson "went cavin.'"

While Knox Cave was closed to the tourist trade after the roller-skating rink burned, it was still visited by the caving community, most of it being "a beginner's cave." Organized cavers removed the old stairway and replaced it with solid, metal stairs.

In 1975, an ice fall in the Knox Cave entrance sinkhole killed one caver and injured another. The cave was closed for several years after that.

Since 1979, the cave and surrounding property has been part of the Knox Cave Preserve owned by the nonprofit Northeastern Cave Conservancy. Access to the property is strictly controlled, and the caves are closed from October 1 to April 30 to protect its population of northern long-eared bats.

1. Secret Caverns: The Night the Cave Burned Down," RoadsideAmerica.Com

The Search for the Garden of Eden

The Legacy of Lester Howe

The legend of Lester Howe's lost Garden of Eden cave, "bigger and better" than his first, continues to confound modern folk historians and the active band of cave explorers who enjoy their sport in the cavern-rich hills of northern Schoharie County.

Extensive efforts to find the purported underground legacy of Lester Howe followed the reopening of his first cave. Several interesting finds were reported and publicized.

The search continues. Since Howe's death in 1888, more than 150 Schoharie County caves have been found and explored. Most of them are small, wet, and physically challenging, and some are barely large enough to crawl into. The exceptions are: VanVliet's Cave, just outside the village of Schoharie; Sitzer's Cave, about two miles east of Howe's Garden of Eden farm property; Secret Caverns, the competitor; and Sinks by the Sugarbush and McFail's Cave, both to the northwest of Howe Caverns. But, except for McFail's, none of them compares with Howe.

Most caves are named for either the landowner or the discoverer. Some of the more colorfully named Schoharie County caves include Cave Disappointment, Featherstonhaugh's Flop, Cave Mistake, Cave 575, the Freight Train and Caboose caves, Peggy's Privy, Tin Can Cave, and the author's favorite, Cave of the Brown Tooth.

Edward A. Rew (1905–1978), the Cobleskill post office employee who accompanied Arthur Van Voris on many of his explorations for *The Lesser Caves of Schoharie County* manuscript, claimed to have found the Garden of Eden by following a mysterious "Finger of Geology." Rew is often referred to as "Colonel Rew," a rank he obtained during World War II serving in the U.S. Army with the intelligence corps in South America.

In 1950 in a letter to Clay Perry, the author of *Underground Empire*, Rew described his 1931 explorations in a Schoharie cave that he took the liberty of renaming "Rew Caverns." Excerpts from the letter, which accompanied an article Perry had authored on the legendary lost cavern for the Cobleskill weekly newspaper, follow:

> The cave I wanted to tell you about, Rew Caverns, is not located anywhere near the recent exploration of the Garden of Eden Cave, although I believe they are one and the same cave . . .
>
> I went many times [to the Howes' former farm property] by myself, and in sitting on the rock ledges, noted that you could see across the valley many places where Howe's Cave had exits on different levels into the valley, all of which pointed in the general direction of the Garden of Eden.
>
> ### Barn Hides Entrance
> I also noted that Mr. Howe had built his farm at the very edge of a cliff and had brought shale rock from across his land a good 200 yards to fill for a barnyard . . . It became my belief that this barnyard hides the entrance to the cavern he spoke of, or at least it was an intention to make people think that it did.

Rew wrote that after talking with elderly local residents:

> I came to the conclusion that Howe was an honest man and that he had undoubtedly found a cavern which was bigger and better that his original explorations. I realized that alone, I could not dig away the shale that made the barnyard.

The Search for the Garden of Eden

At about this time . . . a gentleman from Schoharie came to Arthur [Van Voris] with the story that he had a small cave over on the bank of the Schoharie River and wanted us to come over and look at it. We went into his small cave for a short distance to where we came to a water crawl, the conditions were such that although it was four or five feet wide, the water was eight or ten inches deep and the roof of the cave about the same distance from the water, which would necessitate crawling through and holding the light in your teeth.

For some reason or other, I didn't feel like going through it at that time . . . Late in the Fall of the same year, I was home going over my cave records and on consulting the map, suddenly realized that the show of exits on the north of the Cobleskill Valley made by Howe Caverns, the Garden of Eden, and this small cave on the banks of the Schoharie River all lined up.

I, of course, came to the conclusion that the Cobleskill Valley was geographically a much newer valley than the Schoharie and that originally the caves that are now known as Secret and Howe Caverns extended on through to the Schoharie River and that the digging of the Cobleskill Valley had cut this cave in two . . . that in all probability the small cave we had explored on the Schoharie riverbank was a side passage from the lower end of this cavern.

I went over at once to the cavern and went through the water hole and found a tremendous cavern. I explored this cavern not much over half a mile back of the general direction of the Garden of Eden and was stopped by a cliff which undoubtedly was formerly a waterfall, at least 50 feet high, and which could not be scaled by a person alone. The cavern was so high that the beam of my flashlight could not reach the ceiling.

It's beautiful in formations, the cliff being completely covered with flowstone. One interesting note that in the cavern I found the only crystal formation I have ever seen in the caverns of Schoharie County.

In the close of his letter, Rew explains why he decided to wait nearly twenty years before coming forward with his discovery:

> After this partial exploration, I had the choice of getting help to complete the exploration and thereby letting the world know of my discovery or keeping quiet and respecting the memory of Lester Howe's secret, [he being] probably the only other person to see this cavern. I decided that it was much more fun to keep quiet, letting only Arthur [Van Voris] know that I had made a discovery, without giving him my ideas as to how I gained my entrance.

Rew's Cavern is believed by most to be one and the same as VanVliet's Cave, along the Schoharie Creek. Later explorers, with more sophisticated equipment, have been unable to follow Colonel Rew's lead, and there seems little interest in doing so.

In his 1966 book *Depths of the Earth*, William R. Halliday, director of the Western Speleological Society, found additional claims by Rew to the editor of *The Cobleskill Times*:

> [The Garden of Eden cave] is richer and more spectacular than any hitherto unexplored cavern in Schoharie County or for that matter, any other in the North. I personally and alone, on a secret expedition, discovered and explored it one night in 1931 to the extent of about two miles.
>
> The Finger of Geology points to the Garden of Eden Cave. Read the geology and keep at it, and you'll find it just as I did!" Rew boasted.

Halliday continued: "At various times, Colonel Rew happily planted intentionally puzzling hints. Some may have been red herrings. Often, he referred to the geological structure of Terrace Mountain, southeast across the valley from Howe Caverns . . . He hinted broadly that he had forced a seasonal siphon [a place where the water meets the ceiling of a cave, blocking further passage] in little VanVliet's Cave."

Rew's claims have yet to be substantiated. There was a lot of hope among cavers that the Garden of Eden (or other caves) would be unearthed during the construction of I-88, a four-lane highway through Schoharie County built in the early 1970s. A few small holes were opened in the cliff face near the old Howe property, but nothing significant.

Rew was living in Rogers, Arkansas, at the time the highway was under construction, spending his retirement years working with wood and selling "Colonel Rew's Good Stuff," a wood polish based on an "old English cabinet maker's original formula."

If his story about the cave was in fact a hoax, the Colonel stuck with it, just as Lester Howe had done. During visits to the Cobleskill area to see family, his niece Anne Hendrix of Schoharie remembers him asking, "Has anyone found my cave yet?" as they drove past the location he had described.

The Colonel died in Little Rock in January 1978.

If the lost Garden of Eden Cave was on or near Howe's property of the same name, Sitzer's Cave on Terrace Mountain would be another possible contender. Sitzer's Cave, about two miles northeast from the former Howe farm, was the object of extensive search and speculation during the late 1940s by an Oneonta caving club.

Sitzer's looks promising. In spring and after heavy rains, a powerful underground stream floods from the small entrance, indicating a cave of considerable potential. The stream forms a small ravine visible from the highway.

Sitzer, at the time the owner of the property, told the Tri-County Grotto of the National Speleological Society that he had been in the cavern for some distance and "had seen some remarkable, large chambers." But in or about 1929, Sitzer continued, "an earth shock started a huge landslide that completely buried the entrance . . . with thousands of tons of rock and earth."

The explorers' efforts were publicized by author Clay Perry in a Cobleskill newspaper article. Perry wrote that the group "succeeded in digging and blasting away enough of the landslide to drain a large underground lake to a considerable extent, lowering the level, they say, about 15 feet . . . some of them wormed their way in and found a large chamber and reported another beyond where they could get to, but said the water must be further lowered before more progress could be made."

Colonel Rew must have felt it necessary to weigh in on these unsuccessful efforts. He also wanted it known, and to remind readers if necessary, that he and Arthur Van Voris were responsible for discovery of some of the caves being written about. The Van Voris group, he wrote, "discovered some 22 caves in this region, including those which were in the paper and in Clay Perry's book being called 'original discoveries.'"

In a June 5, 1949, letter to *The Cobleskill Times*, Rew wrote:

> The purpose of this letter is to needle [the Oneonta cavers] by giving them just enough information about "Rew's Cavern" to make them really get down to business and stop playing around. According to the recent article, they have, to date, found two entrances that would have undoubtedly led them to one of the largest caverns in the east, if they had but persevered.
>
> I can say that it is in the general location of the Harold Sitzer farm, and that if they will be diligent, bold and observant of the geological formations, they will have little trouble, in locating the cavern, which they will find to be over five miles in length, complete with lake and waterfall . . .

In closing, Colonel Rew taunts from his Fort Dix, New Jersey, deployment: "That's all the hints I'll give them for now. In wishing them good spelunking, my feelings are mixed, for until they do make an entrance, it's mine, all mine!"

Further explorations in Sitzer's Cave yielded little. A 1966 guide to the caverns of Schoharie County provides the following description: "The cave consists of several rooms that are connected by very wet crawlways. The cave can be entered for 140 feet to a point where the ceiling almost comes to the water. The passage continues and could be followed even further if the water level ever falls, even by a few inches." Later explorations in the 2000s ruled out any further passage; the stream exits the cave at this point through a passage about five inches in height.

There are numerous sinkholes on top of Terrace Mountain, and some small cave entrances visible along the cliff face on both sides of the Schoharie Valley. Cavers still reconnoiter the area from time to time and occasionally make efforts to dig at sites that look most promising. Even if the Garden of Eden is there and it remains hidden, cavers' hopes of discovering any new cave always spring eternal.

Another claim to the Garden of Eden was fired off in 1929 as the first salvo in the battle for the tourist dollar between Howe Caverns, Inc., and its neighbor/competitor, Secret Caverns.

Less than three months after the reopening of Howe Caverns, the former Nameless Cavern was officially re-christened Secret Caverns and opened to the public. In announcements to the press, developer Roger Mallery said that he decided on the name because of the cavern Lester Howe had kept secret. The August 15, 1929, *Cobleskill Times* reported, "It is believed that this is the subterranean wonder which Howe had in mind at that time."

It is possible, but not likely. Less than a subterranean mile to the northeast of Howe Caverns, Mallery's cavern is a tight yet lengthy fissure cave with high domes and a small, active stream. The cavern is young, according to its owners, formed during the great Ice Age. It is not near to the old Howe property, and it just doesn't fit Howe's description of the Garden of Eden cave as "bigger and better."

Local cave entrepreneur D.C. Robinson weighed in for an October 1934 article in the *Altamont Enterprise* uninspiringly titled, "Robinson Describes Capital District Caves." After the reopening of Howe Caverns, the Garden of Eden cave, he wrote:

> . . . was rediscovered and entered at several places but no one has yet traversed its entire length.
>
> The Secret Caverns development is in part of it. The section opened is entirely natural and very beautiful. It shows about 1800

feet of passage, most of which required a great deal of work to make passable for visitors.

The greater part of this cave is undeveloped and is owned by the Robinsons, who control about 30,000 feet of large caverns in that place. It is hoped that this can be opened to the public at some time for it is, as Lester Howe said, a larger, finer cave than Howe Caverns.

Secret Caverns is an extensive cave system, with some intriguing possibilities, though not at the lengths Robinson describes. A narrow, tortuous connection beyond the 100-foot waterfall on the tourist route has been found to link Secret Caverns and Benson's Cave system to the east. The double cave extends in the form of an inverted trident to account for more than 6,200 feet of cave passage. Another 2,500 feet to the south, the underground stream from the Secret Cavern–Benson system emerges in Barytes Cave.

Clay Perry writes in the 1948 *Underground Empire* that one of the three entrances to Benson's Cave was probably recorded as "Minister's Cave" in a 1906 state geological survey. The story, Perry wrote, was that a local minister would enter the cave's large entrance chamber for its pleasing acoustics and would practice his Sunday sermons there. "It is quite a relief from the numerous 'Devil's Dens' and 'Devil's Holes' scattered about the country," wrote Perry.

Benson's Cave is part of a six-acre preserve managed by the nonprofit Northeastern Cave Conservancy. Access to the property is strictly controlled and the caves are closed from October 1 to April 30.

The highway that approaches Howe Caverns and Secret Caverns is lined with billboards. It is easy to understand the commonly asked question, "Are the two caves connected?" The answer is no, and yes.

Barytes Cave is entered through the old entrance to Howe's Cave, now in the former cement quarry of the Penn-Dixie Cement Company.

In 1904, mining broke through a small crawl space in the floor just inside the original Howe's Cave entrance and uncovered a strange phenomenon—two caves, each with its own underground stream, one stream crossing the other and until then unconnected.

Barytes Cave, of which more than a quarter mile has been explored, heads to the north in the direction of the Secret-Benson's cave system. Making a physical connection of the three caves would yield a bigger (longer) cave, but not a better cave. If possible, it would be tight, wet, and physically challenging. Linking the Barytes-Secret-Benson's cave system to the original Howe's Cave would be an impressive cavern by any standard, accounting for an estimated three miles of underground passage.

"Benson's is an important link in the master cave system and watershed which runs from Secret Caverns, through Benson's, Barytes and out into the mines under the Howes Cave quarry," notes the Northeastern Cave Conservancy. "Since the water from the Howe Caverns system drains in the same area of the mines, this could be argued to be the connection between Howe and Secret."

There is another, long-standing rumor of a connection between Howe Caverns and its competitor. A tight, tortuous crawlway, Fat Man's Misery, extends Howe Caverns beyond the Winding Way to the cave's northernmost point, heading in the direction of Secret Caverns. Fat Man's Misery (as this entire section of the cave has come to be known) has been explored to a length of more than 1,800 feet, and it continues on from there. From the Secret Caverns end, it was reported, "Men who discovered and explored Secret Caverns long before it was commercialized claim that there is another long passage somewhere near the entrance that runs all the way through to Howe Caverns . . . This has been walled up by the owners of Secret Caverns."

Howe Caverns, to this day, remains partially unexplored. The passage through Fat Man's Misery, off the commercial tour, has been known since the mid-1800s, and early tours ventured along it by oil lamp as far as the 107-foot-high Great Rotunda. The explored passage continues for about another tortuous 1,000 feet of backbreaking stooping, walking, and crawling through icy 42° water. In most places, the ceiling is less than 4 feet high. Until the 1950s, explorers stopped at this point. On the first modern maps of Howe Caverns, displayed at the lodge entrance for visitors, this passage is referred to ominously as the "Lake of Mystery, On Which No Man Has Ever Sailed."

Beyond the Lake of Mystery is Reynold's River, named for the former head guide at the cave. Reynolds River enters the undeveloped cave at about 800 feet into the passage and soon disappears behind a pile of collapsed limestone ceiling. From there, the passage "goes" (in caver jargon)—becoming increasingly tight. Explorers can stand in places, the walls of the cave pressing against their chests. It continues northward, passing under the fields of Sagendorf farm, toward the Secret Caverns entrance lodge.

It wasn't until 1975 that four caverns guides—the author, Steve Crum, Chris Walsh and John Clark—together with Bob Addis, a former guide, pushed on beyond this point, continuing in the direction of Secret Caverns. Wearing wetsuits and protective clothing, with hardhats and miners' lamps for light, Addis and Crum pushed this undeveloped section of the Winding Way to its reasonable limits. At that point, another 700 feet from the end of Reynold's River, the narrow cave passage is blocked by a deep pool that meets the ceiling. Above it is loose rockfall. But it continues—unexplored, despite attempts to reach the passage's end.

Plotting this point of the cave on the surface, Addis found it lies approximately beneath the crossroads of Sagendorf Corners Road and Caverns Road, about midway between Howe and Secret caverns!

Connecting the upper and lower ends of Howe Caverns to Secret Caverns through the Winding Way (upper) and Barytes Cave (lower) passages would create the Northeast's most extensive cave systems. Mapped, the system would resemble an italicized pound sign, or #.

Another possible Garden of Eden was explored in the late 1950s and it, too, was not far from Howe Caverns. By coincidence, it was owned by Edward and Luala VanNatten. Jim VanNatten, a former Howe Caverns tour guide, nearly lost his life in a freak accident at the cave in 1930.

The colorfully named entrance pit of the Sinks by the Sugarbush is in the dense maple woods (a sugarbush) just northwest of Howe. There, explorers from Albany came upon perhaps the most exciting find in the search for the legendary Garden of Eden—nineteenth-century artifacts believed to be from Lester Howe's collection.

In an article on the Garden of Eden mystery, Clay Perry wrote in *The Cobleskill Times*:

> An astonishing find was made in the VanNatten cavern, a pair of oriental bronze objects which are believed to be Arabian hookahs or water pipes, which are perhaps adapted to use as oil lamps. They bore the crudely scratched initials of "L.H" and the date, "1845." These are believed to be of Indonesian or Arabian origin, dating back before the Christian era, and perhaps brought to the country by some sea captain, who made port at Hudson when it was a whaling port. And that Lester Howe gained possession of them and took them into this cavern, for what purpose can only be surmised.

Gail Hults, whose grandparents owned the busy farm, remembers going to the sinkhole entrance as a young girl when the cavers came. "I wanted to see what was going on."

The cavers were well organized, she said, using walkie-talkies to communicate up and down the pit. "They found a large room, blocked by a wall of clay. They could hear a waterfall beyond it, but they couldn't get through."

According to Hults, the cavers dropped crushed peanut shells in the water. As many suspected, she said, they were later found in the stream that runs through Howe Caverns.

The Sinks by the Sugarbush cavern remains impassable and not many try to enter it anymore. Over time, the tight entrance pit has become plugged with fallen rocks and debris. If a bigger and better cave continues beyond the drop, it cannot be reached without drastic measures or explosives. And the date, 1845, would have been just three years after the discovery of Howe's "first cave." It seems reasonable to assume Howe would have had no reason to plot his subtle revenge at such an early date.

The provenance of the hookahs and other artifacts from the Sinks seems to have been lost. Hults remembers seeing the artifacts in her grandmother's home. She believes they went on to the Smithsonian.

There is some discrepancy in the historical records of the exploration of the caves in the Carlisle Center area, which lies to the northwest of Howe Caverns. Two caves, McFail's and Selleck's, were explored during the 1840s; the initials of T.N. McFail and the date 1844 are near the bottom of the entrance shaft of Selleck's Cave.

McFail's Cave is the only cavern found to date that could be in any respect considered "bigger and better" than Howe's.

Descriptions of McFail's and Selleck's caves from the mid-1800s are confusing. Until the early 1960s, both caves were regarded as pretty but small, each having a small lake just off the entrance pit. But rumors from the turn of the century persisted—a stream passage, indicating more cave, extended northeast/southwest, the stream running southwest in the direction of Howe Caverns! But in which cave was it—McFail's or Selleck's Cave—where McFail's initials were found?

At the run of the twentieth century, a systematic search in the Loessner's Woods area of Carlisle Center led to the discovery and exploration of Cave Disappointment. In the early 1950s, three new vertical caves were found in the same area and named (quite colorfully) after their discoverers— Hanor's Cave, Ack's Shack, and Featherstonhaugh's Flop.

It wasn't until 1960 that a cave matching the description of McFail's was found, only about 200 feet to the northeast of Ack's Shack. It took a great deal of pushing—a caver's term for physically forcing one's body through tight, tortuous cave passages; McFail's Hole was dangerous. The area around the entrance was crumbling and filling with debris. Persistent explorers squeezed through almost one-half mile of treacherous passage from McFail's "Hole"—hardly descriptive of a Garden of Eden.

In 1963, however, Fred Stone of Cornell University crawled—literally with his nose to the ceiling—through a tight, nearly water-filled passage with only about two inches of breathing room. He emerged into the main portion of McFail's Cave, now known to be the largest in the Northeast.

The Search for the Garden of Eden

Continuing explorations in McFail's Cave over the last few decades have yielded more than seven miles of underground passages, most of them large walking passages like the commercialized section of Howe Caverns. The McFail's Hole cave is dangerous for amateurs and in parts unstable. A trespassing explorer lost his life climbing from the entrance pit in 1968, becoming a victim of exposure. The McFail's Hole entrance has collapsed completely; entrance is now through either Ack's Shack or Hall's Hole, the latter discovered in the late 1980s.

If McFail's Cave is the legendary Garden of Eden, it should be considered that Lester Howe would have been a very courageous man—or very foolhardy—when weighing the risks today's explorers have taken to push McFail's to its reputed status as "bigger and better." It is unlikely Howe would have taken similar risks considering his age at the time and the primitive cave exploring tools and methods of the late 1800s—unless, of course, there was another, less difficult entrance into the cave.

Because of its sensitive ecology (McFail's is home to thousands of endangered bats) and its unique place in New York speleology, McFail's Cave is owned and protected by the National Speleological Society. All entrances to the cave are locked and gated. Trespassing signs are posted prominently.

McFail's Cave passes another "test" of the Garden of Eden status. The main cave passage of McFail's follows a northwest/southeast fault line that aligns with the fault-controlled passages of Howe Caverns, less than two impenetrable underground miles apart. The Sinks by the Sugarbush are along the same fault. Line-drawn maps of the passages of Howe Caverns can be said to resemble a right hand with the index finger outstretched. Could this be Colonel Rew's mysterious "Finger of Geology?"

Alas, that finger points in the opposite direction. Following the supposed fault line to the southeast leads to VanVliet's Cave, just as Rew noted. And geologists have surmised that VanVliet's was cut off eons ago from a massive cave corridor (of which Howe and McFail's were a part) by two glacial river valleys and by Terrace Mountain. If the mysterious Finger of Geology points to VanVliet's Cave, as Colonel Rew suggested, there's much more to the cave than modern explorers have been able to crawl their way into. It bears repeating that Howe would have been in his late sixties at the time, and unlikely to go crawling through small, wet, New York caves.

Interest was piqued in the 1970s when construction of I-88 was underway. Explosives blasted the east-west route through the northern section of Schoharie County, and there were a lot of rumors in the cave country about work crews finding gaping holes in the hillsides and of drills being lost in "bottomless pits." Harrison Terk, a former manager of Howe Caverns, reported that workers dumped eighty tons of rock into a large sinkhole on or near the former Howe farm. Additionally, two small caves were exposed just across the valley from the old entrance to Howe's Cave. Located at the top of the escarpment, these two rock holes are clearly visible from the I-88 highway. They go nowhere.

If the Garden of Eden cavern was on Howe's property, the cave is likely gone forever: there are four lanes of highway directly over the former home of the great cave explorer and developer.

Did the Garden of Eden ever exist? Or did Howe, an aged and bitter old man, create a myth to confound those whom he felt had swindled him from his property? Only Howe could say for certain.

It should be noted that for all their seeming permanence, caves do change. Spring floods fill once-open passages with clay, mud, and debris; surface and subsurface erosion takes its toll; and weeds, brush, and vines cover once-visible entrances. The undeveloped Fat Man's Misery section of Howe Caverns is filled with sand and clay from spring flooding almost every year. When the crawlway was reopened in May 1988 and photographed for 1990's *The Remarkable Howe Caverns Story*, eager explorers hauled more than 60 fifty-pound sacks of earth from the crawlway passage, which is only about 30 feet long.

The search for the Garden of Eden cave has continued for well over a century since the death of Lester Howe. It is likely that the search will continue for again as long and new explorers will proclaim its discovery, just as Colonel Rew did.

Perhaps there really is a Garden of Eden Cavern, larger and more magnificent than world-famous Howe Caverns. Perhaps in the future a team of explorers will push their way into a small, tight crawlway that had been previously overlooked or stumble upon a hidden cave entrance somewhere in the deep woods. There, in a vast room miles into the hillside, set among abundant crystal-like formations, the explorers may find chiseled in the limestone wall or written in soot from an oil-burning lamp:

"Garden of Eden Cave. Discovered 1855 by L. Howe."

9

Modern History, 1929–1990

*Lester Howe's Descendants Visit the Modern Cave;
a Brief Geology Lesson on Howe Caverns and Secret Caverns*

By the time of the caverns' reopening in 1929, the Howe family had long since moved from the area. But many of Lester Howe's descendants continued to show interest in the cave over the years, and they maintain a great sense of family pride in its discovery. It is rare when during the course of the year at least one distant relative does not tour Howe Caverns.

Helen Howe, a Broadway showgirl and great-granddaughter of Lester's brother Elmon, married Anthony Gianelli of Schenectady in the Bridal Altar of the caverns in December 1929. It was the fourth cave wedding of the modern era. Four years later, in September 1933, Howe's great-grand-nephew, Herbert Howe of Schenectady, was married in the cave to Rose Gardner of Middleburgh.

For many years, one of the most prized possessions held by the Howe family was the guest register from their Cave House Hotel. The register, which covers the years from 1844 to 1855, was presented in 1965 to the Schoharie County Historical Society by Howe's great-granddaughter.

In written remarks for the presentation, Frances Howe Miller recalled:

> Our grandmother and our mother, Annie Laurie Dewey Miller, both delighted in this book. I recall from childhood the family

dinner parties which we were invited to at our grandmother's home in Jefferson City, Missouri.

She [Harriet Howe Dewey] used to get out what we called her "Cave Book" and read to us the most interesting entries, especially the various poems which gave her particular pleasure.

In a pamphlet history of the Howe family, Warren Howe writes:

A family story, related in 1985 by Helen Howe Gianelli, told of an incident which occurred when a noted scientist, author, and explorer visited the cave. "With great flourish and cameras . . . he made an exploration." At a subsequent college lecture the scientist told of the wonders of Howe's Cave and described how far he had gone into it. "He said it wasn't safe to go farther without scaffolding." Helen's grandfather, then a young boy, laughed at the scientist's claim. He told his father [Elmon Howe], "Why I have been there much farther." He described the "water and bright stones further back in the cave."

In the late summer of 1954, the caverns' general manager, Howard Hall, received a letter from the wife of Charles E. Dewey, Howe's grandson. Her husband's parents were Hiram and Harriet Elgiva (Howe) Dewey, married in the Bridal Chamber in the old cave more than a century earlier. The letter was published on the front page of the *Cobleskill Index* on August 5. The text is printed below. Railroad aficionados may find the letter of particular interest.

Daughter of Lester Howe Was Married in Nearby Caverns a Century Ago

September twenty-seventh of this year will be the one-hundredth anniversary of the marriage of my husband's parents in what was then known as the Bridal Chamber of Howe's Cave. We had hoped to be there for a celebration, but it looks as though we shall not make it. However, I thought as a matter of interest to visitors you would like to choose some items from data which I can supply.

The bride was Elgiva, one of the daughters of Lester Howe, owner of the cave. She was small and retiring with blue eyes and an abundance of light brown hair. The groom, whose name is tucked away inconspicuously in a corner of the invitation to the wedding reception, was six feet tall, handsome, and fun-loving, with dark brown hair and deep blue eyes. However, Mr. Hiram Shipman Dewey was a person of importance even then, being a civil engineer employed in railroad work. He had worked for the Saratoga and Whitehall, the Whitehall and Rutland, the Harlem Extension and later for the Albany and Susquehanna and the Schenectady and Athens railroads. He made the first survey for the West Shore from Hoboken to Newburgh and in 1859 and 1860 he was chief engineer for the New York and Connecticut boundary commission. He made a survey of the entire line, located and erected all the monuments, and the work was confirmed by the legislature of both states, whereas previous surveys had been condemned. In 1865, he went to Kentucky and supervised the building of a hundred and forty miles of railroad. Three years later he went to Springfield, Illinois to be with the Wabash Railroad and in the fall of 1878 located to Jefferson City. He had left construction work in twenty-one states and had spent three years on the Pacific coast.

We visited the cave back in the rubber boot and torch days and regret that we have not been able to enjoy all the wonderful improvements you have made. We congratulate you upon improving and preserving the site and wish you the best of luck always.

Cordially yours,
Anne Stuart Dewey
(Mrs. Charles E.)
Holts Summit, Mo.

PS: Mother went with her daughter back to the scene of her wedding in 1904. My husband remembers visiting there when he was a little boy and seeing his grandfather stand in the boat while crossing the lake and strike the "harp" with his cane.

The one-millionth visitor to Howe Caverns was Caroline Miskimen of Philadelphia, who toured the cave on June 18, 1948. Amid the massive publicity effort, Miskimen was interviewed underground on WGY-Schenectady radio and received an 8,000-year-old stalagmite mounted on a mahogany base with inlaid cave "pearls." She was also presented with a copy of Clay Perry's *Underground Empire* autographed by the author.

Despite its nearly 200-year history, there was little accurate scientific data collected on the famous cave until after the turn of the twentieth century.

Before that time, a distinction between the works of God and the natural sciences was unclear in the public mind. Nineteenth-century scientists had formulated several rudimentary theories on the observable natural processes, and many of these theories were later proved to contain at least some element of scientific truth. But by and large, the American public still held to their belief that the Heavenly Creator was responsible for crafting the great geological works of nature. Advertising material from as late as the 1930s referred to the cave's formation as "a wonderful creation and an interesting study for those who adore the Great Author of the Universe, and delight in contemplating His wonderous works." Yet, as early as 1846, a visit to Howe's Cave could plant the seeds of religious doubt. "There creeps unbidden into the mind a vague and indefinite idea that [the cave's] vast proportions must have required ages on ages, more than 6,000 years, in their formation," wrote one early visitor. He referred to the commonly held theory among believers in the literal truth of the Bible, that the world was only 6,000 years old.

By coincidence it was a religious man, the Reverend Horace Hovey, who published what could be called the first scientific analysis of Howe's Cave. Hovey visited the cave in 1880 with his son Edmund; their findings, along with reports on Mammoth, Luray, and Wyandotte caverns, were published in *Celebrated American Caves*, a first in its field. There with five separate editions overall, from 1882 to 1896. It was the first book available on commercial caves in the United States.

While the reverend's geologic report on Howe's Cave was not richly detailed, Hovey wrote that the cave's underlying limestone bedrock was "easily acted upon by the elements."

The introduction to his chapter on Howe's Cave notes:

> The most massive and prominent rocks in Schoharie County are first, the Water limestones, then the Pentamerus limestone, and above, the Delthyris shale. These all belong to the Helderberg division of the Silurian system . . . these formations . . . are so related to each other as to favor the excavation of deep valleys, flanked by cliffs and mural escarpments, the hills rising by successive terraces to mountainous proportions. Several caves [have] been found in this region.

Howe's Cave, Reverend Hovey concluded, "is one of the largest in the country excavated from the rocks of the Silurian period." (Today's geologists peg the Silurian Period as beginning 443.8 million years ago and lasting a relatively short 24.6 million years.)

Hovey included the first accurate survey and map of Howe's Cave, borrowed from *A Description of Howe's Cave* published in 1865 by Weed and Parsons, Albany. Because portions of the cave have since been destroyed by quarrying, Hovey's map remains the most accurate historic record of Howe Caverns as a complete cave system.

Cave maps are useful tools to speleologists, particularly when used to determine relations with the surrounding surface area. Several maps of Howe Caverns have been updated and revised, often in conjunction with papers presented on the cave geology of Schoharie County and surrounding area. It is particularly useful to determine and predict the underground flow of water and its effect on drainage, erosion, and wells, and to identify potential sources of pollution. In many rural areas, regretfully, sinkholes are often used as dumping grounds for animal carcasses, old vehicles, and other refuse. As a result of these types of studies, the State of New York and Schoharie County have stringent anti-pollution regulations in the cave areas.

The State University of New York at Oneonta is a pioneering institute for the study of hydrology, or the movement of water underground. The SUNY University at Albany is also well regarded for its geology program.

Speleology is a relatively new field of study. The constitution leading to the creation of the National Speleological Society was ratified by cave explorer/author Clay Perry and others in Pettibone Falls Cave in Massachusetts on December 1, 1940.

Howe Caverns, Secret Caverns, and other Schoharie County caves are located in what is referred to by speleologists as an area of "karst topography." In a karst area, surface water drains underground into numerous sinkholes, crevices, and depressions. The Helderberg division, which Hovey described, is a regional name given to the underlying belt of limestone, an easily eroded remnant of an ancient ocean bed that existed more than 443 million years ago.

Geologists say that as sediment accumulated, the ocean floor was compressed into solid rock from the weight of successive deposits, and the continents rose gradually over eons of time. The ancient ocean was filled with life—corals, sponges, and shelled animals similar in many ways to the clams, oysters, and snails of today. Their shells, or armor, were built from calcium carbonate, which is now easily identified as the major element of limestone.

From these characteristics, geologists have identified in Howe Caverns three types of limestone, representing two different periods of time in the history of this ancient sea—the Silurian and Devonian ages. Beneath the hills of Howe Caverns country is a bed of limestone that formed during the Silurian age, as Rev. Hovey identified in the late 1800s, and the Devonian period of nearly 400 million years ago.

The cave, of course, is much younger, formed by the erosive process of water beginning only between 5 and 10 million years ago. The ever-present underground stream, called the River Styx, courses its way southeasterly today through Howe Caverns, having eroded through successive layers of limestone named for their initial geographic locations—Coeymans, Manlius, and Rondout.

Secret Caverns, entirely in the Coeymans limestone, is a younger cave system and more typical of New York State caves than Howe. It was formed during the last 300,000 years of the 2.5-million-year-old Ice Age, when melting water from enormous glaciers found its way below ground to create many caves. The upper layers of the Coeymans bed are rich in fossils. Crinoids, spiral coral, and brachiopods are often pointed out by Secret Caverns' guides as the tour begins the descent into the cave.

Many of Schoharie County's cave systems are carved out of the easily eroded Manlius limestone, in which the major corridors of Howe Caverns are formed. According to the *Schoharie County Guide*, "The Manlius is thick-bedded, dark blue limestone of fairly pure composition. The layers are one to three inches thick, and often alternate dark and light beds. The thin-bedded character of the rock is extremely favorable for the passage of underground water." Manlius limestone is said to be rich in fossils, although fossils are rare in Howe Caverns.

Early visitors to the cave were enthralled with its acoustic characteristics. Of the Manlius limestone, the county guide to caves notes, "When struck with a hammer, it emits a ringing tone."

The ceiling of Howe Caverns is capped by the bluish-grey Coeymans limestone, described by geologists as rather coarse, composed of fossilized fragments of shells, crinoids, and stems. A guide to the area's caves notes, "Although caves in the Coeymans are uncommon, it frequently contains sinkholes and shafts to the Manlius below. The cliffs of the Coeymans are almost everywhere visible in the region."

The original, natural entrance to old Howe's Cave at the caverns' lowest point is carved from the Rondout limestone, "a thin-bedded, drab-colored lime mud rock."

Although these geological features are not described by the caverns' tour guides, astute visitors to the cave today can spot the point of contact between the Coeymans and Manlius limestones as they descend the stairs at the base of the elevators. The Manlius caps the Rondout limestone in the deeper section of the cave, near the boat dock of the underground lake.

The cave—that hollow cavity within the limestone walls—began to form after the Cretaceous Period, some 65 to 136 million years ago, when the North American continent began to rise slowly out of the sea. Rains fell; some drained off in brooks and rivers, and some found its way below ground, seeping into the easily eroded limestone. Taking the path of least resistance, these underground streams followed natural cracks, fault lines, and fissures below the earth.

Speleologists have found evidence in Howe Caverns of two types of water erosion—phreatic, from the Greek *phreat* for "well," and vadose, from the Latin *vadōsus*, meaning "shallow spots." Pockets of water accumulating below the water table slowly erode into domes and chambers through the phreatic process. This explains the formation of domes and other pockmarks in the caverns' ceiling that seem to defy gravity. As the water table lowers, an active stream finds its way underground, then connects these phreatic chambers to form the master cave system. This is known as vadose erosion.

Once the caverns' chambers are open, formations—stalactites, stalagmites, flowstone, and other varieties—begin to develop at the incredibly slow pace of about one cubic inch every 85 to 100 years. Cave formations develop similarly to the way icicles are formed. Rainwater seeps through the soil above, dissolving small amounts of limestone in the process. As these drops of water filter through the caverns' roof, the water evaporates and leaves behind tiny particles of a slightly altered limestone, called calcite, on the rock surface.

The formations hanging from the ceiling are known as stalactites. Stalagmites, the formations that grow up from the cave floor, are formed in the same way. Often they are found directly below stalactites and result from drops that fell to the floor before evaporating. Flowstone is formed in the same manner, but the water evaporates as it flows down the cave walls.

An early view of the caverns' geology was prepared in 1936 for the corporation's *Story of Howe Caverns* by Professor Harold O. Whitnall, former head of the Department of Geology and Geography at nearby Colgate University in Hamilton, Essex County. Professor Whitnall was eloquent in his praise of the caverns' beauty. For visitors, he concluded:

> The memory of the underground world lingers. New conceptions of time and law and beauty are theirs. Inspiration comes to many to learn more of the marvels of geology, but aside from scientific zest and philosophic meditations, there remains for all the dream picture of a crystal palace like unto one that Pluto may have built for his stolen bride, Persephone.

Going Underground

"A Romantic Trip"[1] through Howe;
A No-Frills Trip through Secret Caverns

Under a peaceful hillside
Green with meadow, glade, and tree
Nature's hand has fashioned caverns—
Wonders that we now may see.

Rippling water courses through them
Chambers clothed in shadows deep,
Domes and archways now awakened
From a million years of sleep.

—R.L. Thomson, from *Story of Howe Caverns*,
© 1936 by Howe Caverns, Inc.

Today, Lester Howe would hardly recognize his cave. It is well lit with clean brick walks. The caverns' original entrance now stands alone in an abandoned cement quarry, and about 300 feet of his "old cave" have been destroyed. A 156-foot elevator descent takes visitors through the manmade "back door" of the caverns; tourists see little more than a half mile of the old Howe's Cave.

Secret Caverns, now a third-generation family affair, has changed little—other than its billboards—since it was first opened in 1929.

The public presentation of Howe Caverns is largely the work of local historian Chauncey Rickard of nearby Middleburgh (1874–1938), who wrote the tour guides' spiel for the caverns' reopening in 1929. Drawing upon his knowledge of Greek mythology, Rickard named many of the points of interest in the cave and wrote the fanciful descriptions of them.

Rickard stood well over six feet and weighed 200 pounds plus. He was much admired by those who knew him, and it is unfortunate that by today's standards his original writings seem melodramatic to the point of being humorous.

Imagine the reaction of visitors today to Rickard's description of the underground lake and boat ride, presented in his booming basso profundo voice:

> You have long ago left the familiar world behind. You have brought here nothing from it save memory, and you are now ready to embark on this beautiful fairy sheet of water and ride out into the abode of the Gods of legend and mythology.
>
> I now summon Vestal and Vulcan to come forth with their fires and light the waters of enchantment, the Lagoon of the priestess of beauty, Venus.
>
> You now see the blue light of Vestal, soon to be followed by Vulcan the God of fire who floods the realm with vast spark, as it were, from his ancient anvil. [The boat now leaves the dock.—Ed.] Here on the left is the statue of Memnon, whose armor was made by Vulcan.
>
> We now enter the Hall of Adonis, a choice portion of the lagoon, for here Venus has been lavish with her beauty. It is in the hall of the beloved youth, Adonis, who, when killed in the chase, the goddess sprinkled nectar upon his blood from which flowers sprang up. The ceiling of the abode of Adonis is encrusted with formations typifying the flowers which sprang from this beautiful votive.
>
> On the left is the dark abode of the vanished Gods from which

spirits Zeus has turned his face, and ahead up on the right now appears the mountain and castles of Valhalla, the abode of the ancient warriors who fell on the battlefield fighting bravely. It would seem that here Venus welcomed those warriors from the northland, for the beauty of the world of myth became hers.

Directly ahead you see a massive pillar of Hercules, dividing the lagoon into two parts. At the base of the pillar is the little bay or cove of Alcestis, whom the mighty Hercules brought back from hell.

[The boat now reverses its course.]

On the right is the arm of Hercules, mutilated, yet powerful in its ruin.

Looking ahead, there now appears the Ramparts of the Elysium where repose the spirits of the virtuous Gods. The boat is now entering a series of rocky cloisters. Note the cathedral-like contour of these rocky vaults. Through an almost perfect arch is seen another view, the majestic organ.

Since it was opened, Howe Caverns has provided reliable summer jobs for high school and college students. The guides are mostly young men. It's difficult to imagine a teenager of any generation being comfortable repeating what Rickard had prepared for them, or the recitation being easily understood by visitors who come from all walks of life. But times change.

Rickard was prolific. He even wrote poetry on behalf of his employer.

In the land of the falling waters
Where bride-white birches sway
Is a cavern of rarest beauty
I fain would see some Day.
It has tunneled the rocky hillside
To hide a jeweled stream
For beneath the tall oak and sumac
Its tideless currents gleam.

—from *The Marvelous Howe Caverns* souvenir photo album, published in the early 1930s.

The spiel has become less dramatic over the years, and there are some formations in the cave to which Rickard gave a name but that are no longer part of the one-hour-and-fifteen-minute tour. To Rickard's credit, the names of the caverns' current points of interest have remained largely the same and the story is told only slightly differently. The guides are allowed a good deal of leeway as to how they tell the same tale up to four times a day. They often toss in a joke or two they used many times before, knowing it can get a laugh.

The cave tour is difficult to describe in words alone. Virgil Clymer, the caverns' first general manager and an attorney by trade, took great pains to do so for *The Story of Howe Caverns*. Those planning a visit to the cave may find it helpful; several points of interest, such as *Dante's Inferno*, are described here but are no longer pointed out by guides. Clymer also adds details that go beyond just the visual presentation. Points of interest are italicized.

Clymer wrote:

Howe Caverns is a type of its own—just as Nature made it. It differs radically from other caverns, being a beautiful cavern made two hundred feet below the surface by an underground stream. For more than a mile the visitor winds his way through Caverns high and wide in which he finds a beautiful lake, galleries, and halls. There are few experiences more fascinating than standing in one of the great chambers and viewing the mysterious, beautiful, and colorful stalactites, stalagmites, and rock creations looming like spectral giants or flaring in delicate veils like floating fairies' wings.

Two modern elevators descend one hundred and fifty-six feet, noiselessly and with almost imperceptible motion, from the floor of the lodge to the *Vestibule* of the caverns—a large circular rock-bound room—where the underground trip begins. The purity of the air is at once a subject of comment. The daily temperature is practically uniform at 52° to 56°F, regardless of the extreme of heat or cold at the surface.

Passing through the portal down a gentle incline, the first stalagmites are seen. They are cone-shaped and of a beautiful bronze tint. Next is a view of the *River Styx*, two hundred feet below the Lodge, which during the ages has been cutting its channel through the rock. Attention is directed to a telephone, and the guide advises that telephones throughout the Caverns are connected with the Lodge. Next, one is astonished to see *plant life far underground*. It made its appearance after the lighting installation and includes several species of mosses and liverwort. These tiny growths are pioneers and explorers of the plant world. Next, above the one-hundred-and-fifty-foot bridge which spans the chasm, is *Juliet's Balcony*, a lovely group of clustered tassels which have taken on an ivory tint. The sightseer then is awed by a massive *Balanced Rock* which seems to be lightly supported but has maintained its position for hundreds of centuries. Adjacent to the Balanced Rock is a stalactite resembling a *Hand of Tobacco*, hung up to cure. Next is seen a vast terrace of stalactites, stalagmites, and flowstone, vari-colored from a light cream, up through shades of ivory into rich bronze. Other formations near the spot are the *Fish Market, Flying Boat, Lighthouse* and *Sentinels*.

The Rocky Mountains are an imposing pile of rock, one of the wonders of the subterranean world. The space between the highest crag and the ceiling gives the effect of mountain scenery and a skyline. Below the level of the pathway the rocks are honey colored with grottoes illuminated with floods of crimson light, which suggests the name, *Dante's Inferno*.

Beyond the Rocky Mountains is the highest vaulted ceiling in the caverns, where the mechanical forces have worked through two periods of geological time. This, called the *Temple of Titan*, is a vast elliptical-shaped chamber, fit dwelling place for Titan, the monster of mythology. At the entrance is a huge flowstone formation resembling a *Turtle* standing on its tail. Children like to imagine that it represents some mythological character transformed into the ungainly shape.

The *Chinese Pagoda* is a cylindrical formation of beautiful ivory-tinted calcite. It is eleven feet, six inches high from the bed of the River Styx in which it stands, and nine feet, six inches in circumference near the base.

The *Tower of Pisa* is a leaning cylinder of calcite with water-worn indentations. It recalls to mind the leaning tower under Italy's sunny skies.

The *Witch of the Grottoes* is a profile in projecting rock of an ancient hag with bony features and jagged teeth. A second witch not quite so strikingly outlined leers over her shoulder.

The *Great Beehive* is a stalagmite about twenty feet in height, which resembles on a mammoth scale the object for which it is named. The floor on which this unique formation once rested has been worn away by erosion so that the hive is nearly suspended.

Titan's Fireplace is next viewed with an opening above insuring good draft. This is the "flue." Glowing embers are cunningly suggested by red lamps concealed so that only the color is visible. A little further on a stalagmite known as *Huckleberry Finn's Hat* amuses the observer, and a second *Giant Turtle* "suns" himself. Both are in the bed of the River Styx.

The visitor now approaches the natural island, passes the *Inverted Village* and *Home of the Fairies*. The Inverted Village shows a church steeple with a white cross, upside down; a barricade with a tower; and other oddly shaped, small stalactites. The Home of the Fairies is a mass of colorful flowstone covering the side wall of the cavern below the inverted village. The cavern ceiling over the *Pool of Siloam* is a series of water-worn, curved arches which would do credit to a Phidias [a mythical sculptor]. These arches are extremely fascinating, but their charm is enhanced by the wonderful reflections in the pool. Near the end of the pool are *ancient sun cracks* two hundred feet underground.

Through the next passageway, called the *Cathedral Archway*, is seen the *Pipe Organ*, formed of large, coalesced stalactites, which viewed in perspective resemble organ pipes of ivory hue, beautified with a rich, satiny sheen. An organ note is actually sounding and can be traced to the overflowing waters of the lake far down the caverns. Opposite the pipe organ is an imposing formation known as the *Bishop's Pulpit*. The atmosphere is so truly ecclesiastical that one feels hushed and reverent.

One of the highlights of the tour today, the underground boat ride, wasn't ready for the caverns' opening in 1929. The two-ton flat-bottom boats, named Alcestis and Venus, were assembled in the cave in time for the 1930 tourist season.

A promotional flier, believed to have been written by Rickard and printed prior to the grand opening, detailed a mysterious (and provocative) "Lady of the Lake."

Rickard wrote:

> A stream flows through much of Howe Caverns, at one place forming a natural lake some 800 feet long, and in places 20 feet deep. And here is the "Lady of the Lake," who, clothed in the habiliments of nudity, has Lady Godiva-like, turned her face to the wall. No mortal who has traversed this passage has ever reported that the "Lady" as even so much peeped over her shoulder.

The Lady of the Lake mysteriously disappeared sometime between the caverns' development and the launching of the Alcestis and Venus. The Lady was never made a part of the commercial tour.

A former guide who worked at the cave before the boats were added recalled Rickard passing with his tour group at the head of the underground lake, making a dramatic appeal to the Lady to reveal herself. "And how are you today, m'lady?" Rickard boomed.

"Why I'm just fine, Chauncy. And you?" answered caverns electrician Owen Wallace, who was working on the lake but just out of sight.

The underground boat ride is Howe Caverns' most heavily promoted feature and often its most memorable attraction. In *Story of Howe Caverns*, Clymer wrote of the boat ride: "The acoustics are such as to make even ordinary conversation musical and should one whistle, the reverberations turn the sound into a chorus of silvery flutes.

"It is easy to fancy one's self under a spell of complete enchantment."

Over the years, four boats have been added, and an orderly traffic schedule accommodates three tours in six boats at the same time on the narrow, one-eighth-mile lake.

In 1931, the General Electric Company presented the caverns a unique gift—an electric "moon," which was installed midway up the lake. (The caverns had purchased 24 miles of cable from G.E., then headquartered in Schenectady, to light the tourist route.) In total darkness, visitors would watch the round, yellow-lamp moon rise from the horizon of the lake as colored lights became increasingly luminescent, all controlled from the boat dock by the tour guide. Over the years, the moon became much battered by inexperienced gondoliers in errant two-ton boats on the lake. It was discontinued in 1975, when the original caverns' lighting system was replaced.

Taped music, fed through two large speakers about midway up the lake, replaced the moonrise. Until the tape became warped by humid cave conditions, boat passengers were entertained by a theme piece from Rodger and Hammerstein's *Victory at Sea* and its dramatic crescendo 200 feet below the surface of the earth.

Resuming the trip on land, we return to Clymer's description of the tour from *Story of Howe Caverns*:

> Such formations are seen as the *Giant Epaulet, Golden Cascade, Grottoes of the Naiads* [water fairies], *Bottomless Pit,* and the

Going Underground

Kneeling Camel with its saddle bag. In a rock crevice back of the Bridal Altar is the exquisite little *Lake of the Fairies*, a shallow pool over which many tiny stalactites are forming. Small and delicate as they are, they were old when America was discovered, as the rate of growth of a stalagmite or stalactite is only about a cubic inch in a century.

The *Bridal Altar*, in an impressive setting in the natural *Balcony of Titan's Temple*, is the romantic place where several marriages have been performed. At another spot, in 1854, Lester Howe's daughter was married, that being the first wedding in the caverns. Looking upstream and downstream from the vantage point on the balcony are two very beautiful underground vistas.

Editor's note: Couples who have married in the Bridal Altar since 1929 have taken their vows on a heart-shaped block of nearly pure-white calcite, illuminated from below. The heart is the creation of caverns' contractor Roger Mallery, who in May 1928 wed Margaret May Provost in a part of the cave near the Howe family's bridal chamber.

Back to Clymer's description of the tour:

Crossing the Rocky Mountains again, a *Climbing Lizard*, resembling the Gila Monster is seen. Along the walls of the chasm, under the long bridge, beautiful effects are produced by electric lights. The beauty is greatly enhanced by the sheen on the chasm walls, dampened by the moisture from the rising waters of the River Styx several feet below. Half-way across the bridge is the *Alcove of the Angels*, a deep recess in the rocky walls, one of the most unusual spots in the Caverns. In it are stalactites, stalagmites, and flowstone. Nature's exquisitely beautiful stone handiwork in a variety of colorings from blue-black and dark cream to pure white.

The visitor now enters a unique and curious example of erosion by water called the *Winding Way*, a tortuous series of esses

[S-shaped turns], about five hundred and fifty feet long, from three to six feet wide, and from ten to seventy-five feet high. The sightseer rounds curve after curve in bewildering succession, for the "Way" is so crooked that one seems to change his direction at each step. Much interest is expressed when it is learned that the bricks in the walk are laid lengthways north and south. The formations in the Winding Way show loveliness on a more delicate scale than elsewhere in the caverns. Nature has sculpted the perfect rosebuds on the lowest ledge of a stalactite. A bit of eternal ice called the *Glacier* is set like a keystone in a natural arch. It is the purest specimen of calcite found in the caverns. The *Kissing Bridge* spans the passageway overhead. The *Broken Idol* is next, a pleasing calcite formation sometimes called Niobe, who in mythology became involved in a disastrous love affair and wept herself into stone.

The visitor now enters the *Silent Chamber*. Here is *Pluto's Niche*. In order that one may experience the depths of silence and of darkness, the guide requests quiet for a moment and extinguishes all light. This is the final touch to the awe and wonderment of the pilgrimage. When the lights are turned on the *Stained-Glass Window* is seen. Nature has colored it a beautiful old rose tint, with cross lines in diamond shapes. It is a solid block of calcite, approximately six inches thick and extends across the Winding Way about eight feet above the walk. The visitor ends his caverns trip by retracing his steps through the Winding Way and is carried to the surface in one of the nearby elevators.

Back in the sunlight, the wonderful mountain and valley panorama seen from the Lodge is vested with new charm. The visitor always will retain awe-inspiring memories of this hidden marvel, wrought by the Great Architect of the Universe throughout eons and eons of Time.

Clymer concludes his description of the cave tour with an appropriate selection from the Christian Bible:

"The words of the Psalmist (104:24) take on new meaning: 'O Lord, how manifold are thy works! In wisdom hast thou made them all: the earth is full of thy riches.'"

If a visitor were suddenly transported back in time to Howe's Cave in 1843, it would be relatively easy to identify the points of interest, then and now. The cave, although now brightly lit and with clean walks, has changed little. What have changed are many of the names of the formations and passages, and some in the "old cave" were destroyed by quarrying.

Of all that Clymer described, only the Rocky Mountains and Winding Way still, to this day, have the names christened them by Lester Howe.

Secret Caverns doesn't have a history dating back to 1842 with a worldwide historical record to document or to call upon for romanticized accounts. Operating on a much smaller budget, there have been no poems, hardcover souvenir books, or videos paid for by the Mallery family to promote Secret Caverns.

Of course, the cave hasn't changed since 1929. (In the chapter "The Grand Opening," Secret Caverns was described thoroughly by Arthur Van Voris in his newspaper series on the "lesser caverns" of Schoharie County.) The highlight of the tour today is the 100-foot waterfall found at the far end of the tourist route, about a quarter mile from the entrance. After descending 103 concrete steps, the cave consists of a winding fissure passage along a trickle of an underground stream. From the very first Secret Caverns brochure: "Sometimes the cavern is high and narrow, like the Grand Canyon; sometimes it opens up into large rooms, forty or fifty feet high; sometimes one must stoop or climb around beautiful formations."

Secret Caverns has always promoted itself as a "more natural" attraction than its larger competitor. Although the 50-foot-plus descent into the cave has been eased by concrete steps, and paths and electric lights have been installed, the first brochure notes: "The cavern has been left in its natural state, and sufficient time is given each party to appreciate the consummate skill with which Nature has carved the Thrilling Natural Wonder."

Another brochure from the same period has a little more detail:

> Proceeding from the concrete stairway, competent guides direct parties through the marvelous passages, explaining the geological features and traditions of such special formations, as *The Celestial Dome, Fairy Land, Niagara Falls, The Fountain of Youth*, the *Stalactite Chamber, Home of the Elves, Liberty Bell, Rotunda, Elephant's Head and Tusk, The Cozy Fireplace*, the *Belfry*, the *Beehive Grotto*, the *Suspended Ceiling* and many more. At the *Suspended Ceiling* arm-sized calcite columns connect the ceiling and a lower limestone shelf, the underside of which has been washed away, seeming to defy gravity.

Visitors in the 1930s were welcomed to "Try a refreshing drink of pure wholesome water from the Fountain of Youth, a cold spring found within the depths of the caverns."

The same brochure, oddly noting the cave as "Open June 14, 1930," was likely printed soon after the April 24, 1930, incident at Howe Caverns in which fumes from the adjacent stone quarry killed two men. It makes the point: "Secret Caverns are at all times assured of a free and continuous circulation of pure fresh air through the surface entrance and the airshaft where an exit is being constructed."

The cave's owners were apparently considering a manmade exit at one point; none exists today. From the brochure: "At the lower end of the Caverns, a winding path leads back towards the entrance through interesting and fantastic formations, where an exit is being provided making an easy return to the Cavern Cabin without delay or confusion at either entrance or exit."

The Cavern Cabin, as the brochure described, was still under construction. An illustration shows it to be three stories, and it was planned to accommodate ticket sales, business offices, and a "coffee shoppe and restaurant booths" in a rustic setting.

"The entire trip through the caverns can be made easily and safely," the brochure proclaims, "with no discomfort, as ample time is given and there are no dangerous passages. Hundreds of elderly people as well as children make this trip and feast on the wonders of nature which they see."

At one point in the 1950s, Secret Caverns had its own airstrip, likely the only one in this rural community. The younger Roger Mallery had his pilot's license (as did his father) and, while attending Cornell Law School in Ithaca, would pilot the approximately 100-mile trip home on weekends to look after the family cave.

The most notable changes at Secret Caverns have been made following unforeseen challenges. The cave's psychedelic, tongue-in-cheek billboard campaign, described earlier, began after a 1986 tornado destroyed the many original road signs that took a more conservative advertising approach. On August 22, 1995, a fire of undetermined origin destroyed the entrance lodge. As the Mallerys rebuilt, they took a page from the caverns' billboard designs. Visitors are greeted by a lodge that looks like a bat with its wings spread and an open mouth where the doors are.

Howe Caverns Miscellaneous, 1929–1990

The stream of water that runs through Howe Caverns—the River Styx—continues to rise and fall according to outside weather conditions. Occasionally the cave is closed because of flooding. There is no danger to tourists of being trapped by rising water, but several have gotten their feet wet over the years in hasty retreat to the elevators.

In 1938, a tremendous flood washed out the original gravel walkways. The gravel was replaced by 88,000 bricks as a more permanent path through the cave. During perhaps the worst flood to date, in July 1976, the cave filled to the ceiling with water. The force was so great that it ripped whole sections of brick from the caverns' manmade path.

Howe Caverns and its facilities will never be without electric power. Two large generators in the basement of the lodge supply backup power—enough to light a community of 2,500 homes.

There are few bats in the well-lit portions of Howe Caverns. Approximately 1,000 brown bats live in the cave near the original entrance, which opens into the abandoned cement quarry (see Section III, Chapter 4).

"Howe Caverns," in letters 10 feet high and of white cement, was placed on the hillside at the base of the entrance lodge in the late 1950s. For many travelers, this is their first impression of the site as they approach from any of the roads that lead to the cave. It is probably the most frequently photographed portion of a visit to the famous cave.

Roger Mallery, the former Howe Caverns contractor and developer of Secret Caverns, was first to be married in the cave under the corporation's ownership, in 1928. As noted earlier, he married Margaret May Provost, daughter of the Provosts who ran the old Cave House Hotel where many workers were staying. Since then, hundreds of weddings have been performed in the cavern's Bridal Altar, the couple standing on a translucent block of white calcite carved in the shape of a heart. Permission and special arrangements can be made through the cave's management.

The cave has been used on occasion as the setting or background for film and audio productions. A 1972 television movie, *Tom Sawyer*, was filmed in the cave. It starred young actor Josh Albee as Tom Sawyer, and noted actors Buddy Ebsen and Jane Wyman, with Vic Morrow as Indian Joe. Several Howe Caverns employees had bit parts.

Musicians perform in the cave on special occasions, and some have recorded there, with mixed results. In the early 1970s, the author accompanied an audio production company into the cave after hours. In the still of the cavern, they hoped to isolate and collect the sound of dripping water to use in sound effects. Unfortunately, the microphones were so sensitive they picked up the sound of the cave's ventilating fan, well beyond the end of the tourist route. The audio results were mixed.

Going Underground

In 1979, the corporation celebrated its landmark 50th anniversary with commemorative activities held June 1 and 2. Unique to the occasion was a reunion of the cave's former guides, celebrating "50 Years of Guided Tours." Nearly 200 returned to the site of their summer jobs to enjoy the catered affair. Those attending regaled one another with tall tales and amusing anecdotes, occasionally at the expense of the touring public. Among the long-standing favorites: "Is this real air we're breathing?"; "How many miles of unexplored cave are there?"; "Is it dark down here at night?"; "Has this cave always been here?"

Up until the 1980s, the Howe Caverns guide force was an all-male domain. When generations of guides got together for the reunion, there were numerous tales of amorous guides—all teens at the time—making friends with attractive young ladies on their tour and then stealing a kiss under the "kissing bridge" formation when the lights were turned off.

"We tried to develop techniques whereby we could maneuver some very attractive young girl to the very front of the party, then when the lights were about to go out, say, 'step right down close to get a better view . . . ,'" recalled Harold "Bud" Jones, who worked as a guide in the 1930s. But times and sensibilities have changed, and he cautioned against such hijinks today. Jones shared his memories, which he called "The Howe Caverns Years," in 1997 in *Windmill Village Memoirs*, a retirement community publication.

An anniversary program booklet was assembled, and letters of congratulations were solicited and began pouring in. A resolution honoring the guides' "enthusiasm and concern" when "encouraging . . . an appreciation of our nation's natural wonders" was received from President Carter's Secretary of Commerce, Juanita Kreps. Similar honoraria were presented by the New York State Assembly and the local Board of Supervisors.

Since 1842, Howe's Cave has captured the imagination of countless millions of visitors. Lester Howe's discovery took place at a time when the American public yearned to know more about the ever-expanding world around them and eagerly sought their country's natural wonders. They were inspired by the early naturalists, such as Schoharie's father-and-son team, the John Gebhards, and by the period's popular literature, painters, and explorers. The mysteries revealed in Howe's Cave satisfied a growing public demand and created a profitable—if unusual—business. It became a very human and dramatic story, with ups and downs, life and loss, love and heartbreak. The history of Howe's Cave unfolds like a plot in a Shakespearean play.

Visitors to the cave of over a century ago endured hardships that few nature-loving travelers would tolerate today. Led by crude torches and oil lanterns, the curiosity seekers of the nineteenth century spent hour upon hour in the cave. They emerged cold, wet, and caked with mud, and in most instances grateful for the novel experience. It inspired them to write poetry and compose lengthy letters to friends.

Throughout the mid- to late 1800s, the cave's beauty was greatly magnified, its size grossly exaggerated. (It is likely that the myths surrounding Howe's Cave developed unintentionally, fed by a public need for grandeur in nature.) Scientific inquiries and methods were refined as the century got older, perhaps removing some of the "mystery" of the cave's natural beauty as perceived by visitors.

Those associated with the cave suffered as well. Lester Howe, discoverer and eager host to the scientific minds of the day, was victimized by individuals who represented the changing times. Coinciding with Howe's fall from grace in the public's mind was the embrace of the new symbol of American greatness—the aggressive giants of industry. Joseph Ramsey—politician, railroad tycoon, banker, cave owner, and manufacturer of cement—was a respected and admired leader of the early industrial revolution in New York State.

The age of industry, represented by men like Ramsey, nearly destroyed Howe's magnificent cave by exploiting its resources for their utilitarian value in the building trades. Left unfettered, perhaps they would have succeeded.

But it was the same industrial spirit for innovation and technological advance that rescued the cave with modern engineering techniques not possible until the early twentieth century. Recall how impressive it was

Going Underground

thought in 1929 to have strung telephone lines 200 feet below ground to communicate with the lodge!

Tourists—100,000 or more each year—still marvel at the caverns' beauty, but it's not so easy to appreciate the rarely told human drama of the cave's nearly 200-year history. There are countless stories behind the well-kept brick paths, attractive lighting, and comfortable stroll 200 feet below the surface of the earth. Many of the personalities—and some of the tall tales—associated with the long history of the cave have endured. Most notable of course is Lester Howe, the "eccentric genius," whose place in the annals of New York State and in the history of cave exploration is permanently assured. Stories of Lester's odd behavior continue to be shared, and ambitious cave explorers still search the hills of Schoharie County for the fabled Garden of Eden Cave, "bigger and better." The other families and individuals whose lives have affected and been affected by the caves are lesser known, undeservedly so. The Ramseys, John Mosner, the Sagendorfs, the Mallerys, the Robinsons, the VanNattens, Edward Rew, Arthur Van Voris, and Clay Perry, as well as countless others, have made the story of Howe Caverns the most unique in the history of American caves and cave exploring.

---

1. From *Story of Howe Caverns*, © 1936 by Howe Caverns, Inc.

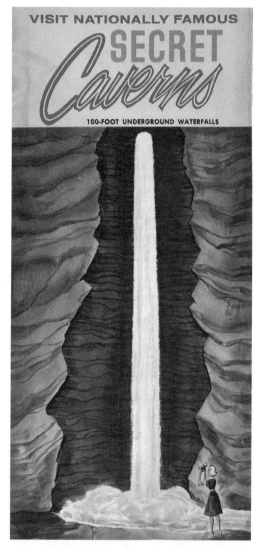

Secret Caverns trifold brochure cover from the late 1960s.

SECTION II

Unearthing Howes Cave
The Community and the Quarry from 1842 On

A Famous Discovery

More than Just a Cave

The residents of Schoharie County in central New York have always known they had unique and plentiful natural resources to draw upon. Farming was and is the primary source of income. A growing number of families settled the fertile valley on New York's frontier following the Revolutionary War's end in 1783, and population more than tripled in Schoharie County by the 1840s.

The county was created in 1795 by joining portions of Otsego and Albany counties. The limestone "cave country" is approximately the northern third of this 622-square-mile rural area; the better land for farming was found in the lower two-thirds through which the Schoharie Creek winds its way.

It was still possible to farm the fertile northern third of the county, but the land is undulating and rocky. Some lands contain depressions, sinkholes, and caves—or all three. Most farmers of the early 1800s had little or no use for sinkholes or caves. A lucky landowner might find a cave entrance large enough to store some perishables over the winter; another might find a cave spring, providing an inexhaustible source of water that could be diverted for irrigation purposes. These finds were rare, however.

One lucky farmer, though, hit the jackpot in the mid-1800s when he purchased a rocky, sinkhole-pocked piece of land north of Howes Cave in the Town of Carlisle. ". . . located on what was formerly known as the John Russell farm and later as the McMillen farm, there is a distinct

A Famous Discovery

Four-year-old Julia Fullerton stands on the footbridge that once crossed the Cobleskill Creek from the Howes Cave quarry to a small community of houses built for quarry employees. The photo was taken in 1916.

phenomenon of nature unknown to many in these parts. This phenomenon is the Ice Cave," wrote Arthur Van Voris, at the time 88 years old, in the summer 1978 edition of *The Northeastern Caver*.

Ice Caves are quite rare.

Van Voris recalled the 1934 interview he had with W.E. Angle, the 81-year-old historian of Carlisle: "He said that fifty years before, it [the Ice Cave] supplied most of the townspeople in the entire countryside with their ice throughout the summer months."

Angle told Van Voris how he and his father in the 1870s would drive their horse-drawn wagon to the cave, hitch the team, and then each, with an ax and basket in hand, would descend the sharp and winding path that led down into the cave.

"They chopped out and loaded on all they needed, for cooling their milk, for preserving their meats, for making homemade ice cream, and the like. And, strangest of all, the supply always seemed to replenish itself . . . right in the midst of summer," wrote Van Voris.

The ice was plentiful enough that the cave's generous landowner made it available to all in the small farming community. "It was customary," according to the historian, "to draw ice in accord with the needs and demand of each farm or household."

The Carlisle Ice Cave is a small cave. The walk-in entrance at the bottom of a deep, sloping sinkhole leads to a room 25 feet high and more than 65 feet long.

Chuck Porter, the long-time editor of *The Northeaster Caver*, added details for the 2008 Town of Carlisle Bicentennial booklet. "The entrance to this cave was bulldozed shut in the late 1970s. A new owner recently began digging it out but has apparently not succeeded in restoring the original Ice Cave."

He explained that the peculiar properties of the cave are due to the rate at which heat travels through the surface of the earth. "Soil and rock are a fairly good insulator, at least on the scale of the 30-foot depth of the Carlisle Ice Cave; temperature changes travel slowly through the ground. Since it takes several months for the cold temperatures of winter to reach cave depths, entering surface moisture freezes in the cave during the spring and summer. By the next winter, the heat of the preceding summer reaches the cave, and so no ice forms then."

Carlisle historian Angle told Van Voris: "Don't ask me to explain it. I can't. It's just there and that's all there is to it!

"Anyhow, that's the way it has been doing since 1850, when I was a boy, and I don't know how much longer before that."

The Carlisle Ice Cave was described in *Glacières; or, Freezing Caverns*, a book of examples of this unique phenomenon worldwide by author and mountaineer Edwin Swift Balch, published in Philadelphia, 1900.

"The north part of the county is mostly underlaid with limestone, which supplies an abundance of good building materials; and as it contains numerous fossils, some of which are very rare," wrote historian Jeptha Simms in his 1845 *History of Schoharie County and Border Wars*, "it affords the practical geologist a good opportunity to investigate this useful science."

There were few "practical geologists" in that period, although it was becoming a burgeoning field, helped in part by two early pioneers, Attorney John Gebhard and his illegitimate son, John Gebhard Jr., both of Schoharie village.

By the time Simms began compiling his history, Gebhard Sr. had discovered and explored several caves in and around Schoharie. His son had identified and amassed a huge collection of minerals, fossils, and other natural curiosities from the area. This collection eventually became a starting point for the New York State Museum. The Gebhards' story is told more fully in Section IV of this book.

Simms's *History* gives an exhaustive account of the first explorations of Ball's Cave (sometimes referred to as Gebhard's Cave) by John Sr. and others, and he describes other caves, such as Selleck's, in the Town of Carlisle. Simms gives a detailed, yet less descriptive, account of Howe's Cave. At the time, Howe's Cave had only recently been discovered. Simms wrote: "Otsgaragee Cavern, known in its vicinity as Howe's cave, and called by E.F. Yates, Esq. [a naturalist and early visitor] The Great Gallery Cave..."

Howe's Cave was named for Lester Howe, the farmer who discovered and made famous the cave that still bears his name. He began to explore the underground wonder on May 22, 1842, after uncovering its hidden entrance on the hillside near his farm.

That tale has been told and retold daily at Howe Caverns since 1929. For purposes here, suffice it to say that Lester, his wife Lucinda, and three children ran Howe's Cave as a popular tourist destination for nearly 30 years, attracting visitors from around the world.

The arrival of the Albany and Susquehanna Railroad in 1869 was a significant milestone in the history of the community that was growing up around the cave property. The Albany-to-Binghamton line included a local depot at which trains arrived daily. It was just a short walk from there to the Howe's Cave House Hotel.

Construction of the rail line began in the early 1850s, and as surveyors and construction crews worked their way past Howe's Cave, they found one of the richest beds of limestone in the Northeast, just a short distance to the west of the Cave House Hotel. By mapping the layers of rock formations in the slope below the cave, geologists found it to be the same as that being mined for natural cement in the Rosendale/Esopus/Rondout area of the lower Hudson Valley.

First came the Howe's Cave Lime and Cement Company, formed in 1867 by Town Justice John Westover of Richmondville, Schoharie County, and Jared Goodyear, E.R. Ford, and Harvey Barker, all from the Oneonta area, Otsego County.

The company capitalized at $100,000 and purchased 70 acres of property that was described as "rude and rugged in the extreme" but included "generous ledges of limestone that promised a good income for a company ready to quarry and process it," according to another early historian, William Roscoe.

Eli Rose (1840–1919) of Richmondville had worked on the family farm and taught school for two years when, at age 27, he joined the new lime and cement company as bookkeeper. The following year he purchased an interest in the business and rose quickly through the ranks, first to foreman, then general manager, and then secretary/treasurer. (In addition to managing the lime and cement business, the ambitious Rose also owned and operated a general store that he started in 1868.)

Rose oversaw the construction of virtually all aspects of the plant, including a pond and a water-powered mill that ground limestone to a powder. Housing was provided for the employees, and barns for the workhorses and mules. Shops, kilns, and mills were erected, and derricks, engines, and other equipment were brought to the expanding Howes Cave mining site. As many as 80 men were employed by the growing company.

About a dozen modest homes were built for employees just across the Cobleskill Creek to the south of the mill and kiln. Over the years, this company housing project took on a life all its own. The locals called it "Tite Nippen"—a name no one can seem to locate the origin of or define in any language. More on Tite Nippen later.

In a biography of Rose written in 1899 for *Life Sketches of Leading Citizens of Greene, Schoharie and Schenectady Counties*, the author opines:

> Fortunately for the company, the line of the Albany & Susquehanna . . . ran near—so near, in fact, that often in blasting, large pieces of rock were thrown on the track.
>
> The ledge nearest the railroad, which is of dark blue limestone, is forty-four feet thick, and is composed . . . of comparatively thin and light rock. Next above this is a ridge of gray limestone in massive blocks and of excellent quality and soundness, such as are eminently suitable for the construction of piers, abutments, canal locks, retaining walls, and all kinds of massive masonry.

[The author here refers first to the Manlius limestone, capped above by the Coeymans limestone. At that time, they were likely cut for building stone or burned and ground as lime mortar. The Rondout limestone used for Rosendale or natural cement was taken from the underground mine.]

The *Life Sketch* for Rose continues:

The lime produced in the kilns is very strong, adhesive, and of great durability. Its lasting virtue is well shown in the stone fort at Schoharie Court House,[1] which was built more than a hundred years ago, and as yet presents no imperfection of either stone or mortar. Among the important structures in which this cement has been employed are the following: the new capitol at Albany; Holland House, New York City; the Scranton Steel Works; Troy Steel and Iron Works; and the reservoir at Fair Haven, Vt.

As all the process of manufacture and the disposal of the output was under Mr. Rose's supervision until his recent retirement [February 1898], no further commentary upon his ability both as an executive officer and as a financier is needed.

From about 1865 to 1890, an underground mine was worked to provide a silty-clayey variety of limestone for natural cement manufacture. The underground mine grew in size over the years; today it lies below

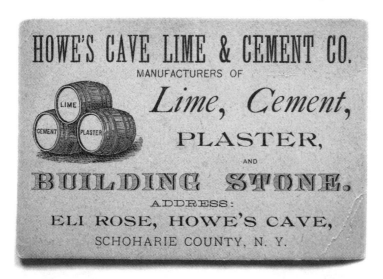

An advertising circular from the first company to quarry limestone in Howes Cave. From about 1870.

about one-third of the current quarry property and even extends below the Howes Cave community.

The underground mine today is closed for safety reasons, but the mud in the mine preserves the ore cart tracks and the hoof prints of the mules that pulled the ore carts. The mules would not work if their ears touched the ceiling of the mine, so quarrymen took out a seventh foot of ceiling to elevate the head room for the mules' behalf.

All that remains of the Howe's Cave Lime and Cement Company today is the foundation of the old vertical kiln, about three stories high and beautifully made of laid-up stone. Its entrance is beneath a gothic stone archway that has stood for well over a century.

In 1869 a second company, the Howes Cave Association, began producing Ramsey's "Hydraulic Cement" and "Wood Burned Lime." Joseph H. Ramsey of Lawyersville was president of the new association, president of the Albany and Susquehanna Railroad and—among other lifetime distinctions—served terms in the New York State Assembly and Senate.

Ramsey (1816–1894) was one of two boys in a family of ten children born in the town of Sharon, Schoharie County. An embarrassingly flattering *Noted living Albanians and state officials* from 1891 noted, "His ancestry is of German and English origin, the more sturdy and substantial qualities of which he has combined in an eminent degree. His father, the Rev. Frederick Ramsey, was a man of high moral and religious character . . ."

The Howes Cave Association quarried the area immediately north of the first company. Roscoe's 1882 history documented it as being "upon the side of the limerock hill, in which Nature has placed treasures which the genius of man requires to aid in the construction of his enterprise."

At first, both firms produced natural, or Rosendale, cement in addition to lime and cut building stone.

Natural cement is a "hydraulic cement"—meaning it sets with water—made from limestone that has high clay content. It is different from lime, which is made from limestone with lower clay content and does not set with water. Both were manufactured and sold at the early Howes Cave quarries.

The process of manufacturing a binding agent like lime and cement dates to ancient Greece. It is a labor-intensive process that at first lasted several days. Strong backs and arms are required.

After blasting from the quarry face, limestone rubble would be broken up by air drills and sledgehammers into smaller pieces weighing about 20 pounds. The stones would be fed into the kilns—from the top—to form layers of about a foot thick with alternating layers of coal or wood. This was called "charging," and it was initially done manually. Tons of stone were fed into the kiln one shovelful at a time.

Built of masonry and lined with fire-resistant brick, the insides of the kilns were between 8 and 10 feet in diameter. They reached temperatures of between 1,600° and 2,000° Fahrenheit to reduce the limestone to "quicklime."[2] It often took several days, and a kiln supervisor had to keep careful watch to maintain the proper temperature. Fumes from the kilns, nearly always burning, contained carbon monoxide.

After burning, workers shoveled the burned quicklime that fell through a grate at the bottom of the kiln. To produce lime, this quicklime would be mixed with water to make it "slake," or crumble to a fine particle size. (When hydrated lime "putty" is used in masonry, it sets by reacting with

Workers in the Howes Cave Association mine in 1889. The photo is part of a series by noted Adirondack Mountains photographer S.R. Stoddard.

carbon dioxide in the air. This is a slow process, often requiring weeks or months to fully set.)

Natural cement—with the high clay content—does not slake when mixed with water. Instead, it must be crushed into a powder before use. Both Howes Cave plants had mills for grinding this powder. The Ramsey plant had two mills, one powered by steam, one by water. The burned limestone passed through "rotary crackers" to further reduce its size before being run through a millstone to become a fine powder.

---

1. This is quite confusing. "Schoharie Court House" was often used to geographically pinpoint the village of Schoharie. The stone fort seems to be a reference to the "Old Stone Fort," a Revolutionary War–era fort constructed a century previously, and not from Howes Cave limestone or cement.

2. Quicklime has many uses. While jailed in Britain, Irish poet/playwright Oscar Wilde watched it being shoveled into graves to decompose bodies more quickly. He composed the following lines:

> And all the while the burning lime
> Eats flesh and bone away
> It eats the brittle bone by night
> And the soft flesh by the day
> It eats the flesh and bone by turns
> But eats the heart away.

The Industry Grows

The Mine Moves above Ground; the Cave House Burns

With a superior product to sell, the cement and stone industry in Howes Cave quickly prospered. In the late 1800s the two companies produced about 200 barrels of cement per day. They also sold building stone, crushed stone, and cut stone of all descriptions. History reports that lime and cement were shipped in that period via the convenient rail line as far west as Indianapolis.

A testimonial from a sales circular of the era:

> I have used 50 carloads of Howe's Cave Cement on the works of the Lackawanna County Court House and have found it perfectly satisfactory.
>
> While it is somewhat slower in setting than some other cements I have used, it is more uniform and reliable than any of them, especially when used under water, and I can cheerfully recommend it as a first-rate cement.
>
> —John Snaith, contractor, Ithaca, N.Y.

Here's another:

> We have been using Hydraulic Cement, manufactured from the Howe's Cave Hydraulic Limestone, for a year past, in masonry on

The Industry Grows

the line of the Albany and Susquehanna Railroad, and find it equal in every respect to the best quality procured from other sections of the state.

We have used it exclusively in the arching of the Webster Tunnel, where the best quality is required, and it has given entire satisfaction.

I have no hesitation in pronouncing it fully equal to the Rosendale or Western Cements.

—C.W. Wentz, chief engineer, Delaware & Hudson Canal Corp.

Ramsey's Howes Cave Association quickly added to their holdings by purchasing an additional 800 acres surrounding their quarry, including the cave property from the Howe family and their Cave House Hotel. (History records it as a transaction of dubious ethics, with Howe accepting 12,000 shares of stock in a new Ramsey corporation after turning down a $10,000 cash offer. See Section I, Chapter 3.)

In about 1872, the association began construction of the third Cave House Hotel, this time of cut stone from the quarry. Soon after, they began construction of a huge, three-story addition that more than doubled the stone hotel's size. Ramsey renamed it the Pavilion Hotel, and its amenities rivaled those of any of the Catskills' more famous resorts. Rates were $2.50 per day; $10 to $15 per week.

The community around the quarry was growing as well. In an 1882 history of Schoharie County, author William E. Roscoe noted, "Brayman's Mills . . . is closely connected with Howe's Cave, around which has sprung up quite a settlement, the citizens of the two places and surrounding neighborhood erected a Reformed Church in 1875."

Helderberg Cement was formed in 1898 by a group of men from Albany who purchased the two existing Howes Cave plants and combined them into one. Joseph Ramsey had died in 1894; his son Charles managed the Howes Cave Association and stayed on as an officer with the Helderberg Company. Eli Rose sold his share in the Howes Cave Lime and Cement Company and retired to Central Bridge, a hamlet a few miles away.

The new offices were moved into the grand Cave House Pavilion Hotel, and while tours of the adjacent cave were still available, they were discouraged. It became more important and profitable to make cement.

With renovations to the plant, Helderberg produced the relatively new mix of Portland cement, first manufactured in the U.S. in 1871. By the turn of the century, the Howes Cave plant was producing 1,000 barrels of this new Portland cement per day.

Portland cement had the advantage of being able to harden underwater.

Manufacturing Portland cement allowed stone to be taken from the higher, more "pure" limestone deposits—in the Manlius and Coeymans formations—at Howes Cave. It meant the mines beneath much of the quarry could be abandoned.

(Manufacturing Portland cement on an industrial scale only became viable following the invention in 1869 of large-scale rotary kilns. It is

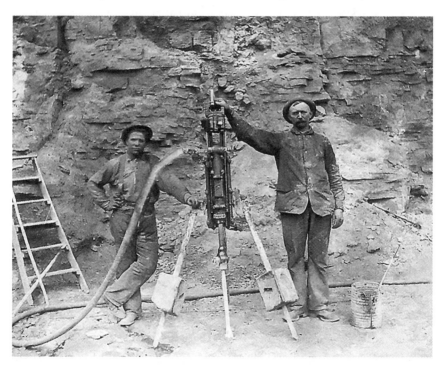

Workers with an air drill in the Portland cement (surface) quarry in Howes Cave in the last decade of the 1800s.

The Industry Grows

named for its similar appearance, when set, to stone found on the Isle of Portland, in the English Channel.)

There are only two remaining Portland cement plants operating today in New York—the Lehigh Cement plant in Glens Falls and the Holcim-Lafarge plant in Ravena, along the Hudson River. About a third of all Portland cement used on the east coast is imported—from Canada, the Caribbean, Central America, Spain and elsewhere.

As a young boy, Fernando Boreali worked with explosives in dangerous, uncomfortable conditions in a marble mine in northern Italy. Immigrating to the U.S. in 1912, he was sent by immigration officials at Ellis Island to work in Schenectady, a growing manufacturing center at the time. Soon after, he found work in the Howes Cave quarry with the Helderberg Cement Company. He had the unenviable job of setting the dynamite charges that would explode 4 to 5 tons of limestone from the quarry face.

First-generation Italian immigrant Fernando Boreali, blurry but third from right, with Howes Cave quarry workers in the early 1900s.

Boreali was lowered down the quarry face in a parachute-like seat and harness to place and rig explosives. Fortunately, he was good at his job.

"If a charge didn't go off, he'd have to go back down and see what the problem was," his son, also named Fernando ("Fred"), recalled in 2005. Now deceased, "Fred" Boreali ran a successful restaurant only a few miles from Howe Caverns and the Howes Cave quarry for several decades. He remembers seeing his father coming home from work covered in cement dust.

A lot of the quarry work required strong arms and backs.

After a blast—which could knock windows out 4 or 5 miles away—the oversize blocks of stone would be broken down manually by air-powered drills and sledgehammers before being placed in the "dinky," a reduced-scale train that would haul stone to the crushing unit.

A photo postcard collection created in the early twentieth century showed a variety of uses to which Helderberg-brand Portland cement could be put. The product had been used to construct the General Electric Building #46 in Schenectady; a circular, hollow-wall concrete silo for farmer Thomas Murray in Amsterdam, N.Y.; the Union College, Schenectady, gymnasium; concrete sewer and water pipes for the City of Albany, and a reinforced concrete bridge over the canal in Cohoes, Albany County.

Produced for the company's sales office at 78 State Street in Albany, the postcard collection also boasted of Helderberg cement being used for the Albany Barge Canal Terminal, which had impressive statistics. The concrete dock was 1,510 feet long and supported by 100 concrete piles.

The company's proximity to General Electric in Schenectady (then in its glory days), as well as to New York's capital in Albany helped Helderberg Cement Company join the ranks of the region's early industrial giants. As such, it was able to attract for its board of directors some of the era's most prominent business leaders and investors.

Historical business records are difficult to come by, but one member of the Helderberg board was Anthony N. Brady, listed in a Yale University archive as a "traction magnate." Brady, a business partner of Thomas Edison, accumulated great wealth after meager beginnings as a clerk in Albany. In addition to sitting on the Helderberg company's board, he served as director/president/chairman of numerous companies nationwide, mostly in the gas, electric and railroad fields.

In February 1900 the huge Pavilion Hotel burned to the ground in a fire of undetermined origin, although the hotel and cave were lit by gas. Only the stone Cave House was left. Because tours of the cave were no longer encouraged, the owners decided against rebuilding the Pavilion, and the Cave House was again renovated to accommodate the cement company offices and laboratory.

It would be another 20 years before Howe's Cave could be seen by curious visitors, for a fee—this time, as Howe Caverns.

An early kiln at Howes Cave that burned limestone at 2,700-plus degrees.

While the cave was closed to the tourist trade, the boys and girls of the surrounding farm country found it endlessly fascinating. Perhaps they'd read *Treasure Island* or similar adventures of buried treasure, and the caves and abandoned mine entrances were great places to act out their daring exploits.

Delevan "D.C." Robinson (1885–1960) is a familiar name in the story of the area's caves. He lived with his family on farm property just up the hill from the Howe's Cave hamlet. An enterprising young man and a tall, broad-shoulder farm boy, he would—for a price—lower younger boys and curious explorers on a rope into the "rock holes" on the family farm, recalled his niece, 96-year-old Helena Ackley, in 2020.

"I think all explorations by anyone [in old Howe's Cave] were free. But there was a lake, and Dellie had a boat in there and probably made a paying enterprise of it, if that was possible," she wrote in an e-mail to the author.

Robinson went on to open his own cave, Knox Cave, in Albany County, as described in Section I.

The gymnasium on the Union College, Schenectady, campus, made with Helderberg-brand cement from the Howes Cave quarry in the early 1900s. From an advertising brochure of the era.

Boom Times Ahead

A Two-Room Schoolhouse, Two Churches, and a Chevy Dealership

With the cement works fueling the local economy, the community around it continued to grow. Photos from the first few decades of the twentieth century are rare, and those that have survived are mostly of the quarry. The few photos available show a handful of homes clustered about the gently sloping hill that leads to the stone works. The Cave House, situated midway up the hill, overlooks many. Other photos from the era look across the Cobleskill Creek from Tite Nippen, where workers' homes were linked to their jobs by two footbridges across the stream.

Ownership of the cement quarry again changed in 1925, when the Security Cement Company of Hagerstown, Maryland, joined with the Berkeley Lime Company of West Virginia to purchase the Helderberg plant and collectively create the North American Cement Company.

North American ran the plant perhaps more successfully than any owner before or after, and the quarry and the community continued to grow and prosper over the next 40 years. Initial improvements, including the addition of a third rotary kiln, made the Howes Cave cement plant the most efficient, productive, and cost-effective plant in the industry. Capacity during the 1920s first reached a high of 3,000 barrels, and soon after, 4,000 barrels per day.

In 1926, the addition of the plant's own generating station to power the quarry operation had unintended benefits for the young people of Howes Cave and their neighbors in Tite Nippen.

Excess heat from the powerhouse was released through a series of pipes suspended over two cooling ponds along the Cobleskill Creek. This hot water exhaust warmed the ponds throughout the year, and kids from the neighborhood would swim and bathe in the ponds year-round.

"Boy, the EPA—Environmental Protection Agency—would have had a wonderful time back then," exclaimed Bob DeRuvo of Aston, Tennessee, a former Tite Nippen resident. Topics like toxicity and carcinogens in industrial operations weren't discussed much back then.

Occasionally the superintendent of the powerhouse would step outside to chase the kids away. "They'd hide in the bushes until he went back inside and then go back to swimming," remembered Tony Spenello Jr., of Cobleskill.

(The power station was shut down in 1955 after the Syracuse-based Niagara Mohawk Power Company convinced North American it could

Another view of the Cave House, in the 1930s. For a few years this was the Provost family's boardinghouse. Many of the contractors working to reopen Howe's Cave stayed here. The entrance to the cave is just to the left, out of frame.

deliver power to the plant more efficiently and for less cost. It took nearly 70,000 volts to run the plant, according to a 1965 report.)

The quarry and community continued to move forward during the 1930s, although the Great Depression took its toll and production slowed to a limited basis. In some years, the North American plant had only to be open a few months to meet the reduced demand.

John Pangman was born in a two-story, two-family home just off Main Street in Howes Cave in 1931 to Donald and Pat Pangman. His dad worked odd jobs at the cement plant when it was running during those years.

"He was very handy, very mechanical," John remembers. "He'd work as a carpenter and welder, earning a dollar a day." His mother sewed baseballs for the major leagues to earn a little spending money.

Despite the hard times, John remembers growing up in Howes Cave with fondness. He was baptized at the Catholic church and attended classes at the two-room schoolhouse.

"There were a number of families that had a lot of kids," he said. "Everyone knew their neighbors and it was a progressive little village to grow up in." The community baseball diamond and children's playground and picnic area were immediately behind the Pangman home. The Depression couldn't break Howes Cave.

In the company newsletter, "Safety Flashes," North American President John Porter told employees the firm had weathered those "rough and stormy days," and in 1942 proclaimed, "Times have changed, our plants are running at full capacity, jobs are secure, and many of our employees are earning more than ever before."

During World War II, the company found its own niche from which to contribute to the war effort. By making minor changes in the manufacturing process, North American was able to process bauxite, the key ingredient in aluminum. Found mostly in tropical climates, the mineral was already being used by the company as part of the mixture that comprises Portland cement. They just ordered more of it.

"The Howes Cave Plant is doing its part to support the war effort," proclaimed the company newsletter in late 1943. In ceremonies on the Fourth of

A busy Saturday night at the Howes Cave Hotel, a popular watering hole for quarry workers.

July, the local chapter of the American Red Cross dedicated a "service flag" to honor the sixteen "boys" from the community serving in the Armed Forces.

"The men from the plant reconstructed an old boiler tube, and made a very good-looking flag staff," said the "Safety Flashes" editor.

(In addition to company news from North American's three quarries, the newsletter included safety suggestions, plant improvements, and personnel items, such as the following from Howes Cave: "Anthony Spenello, kiln supervisor, sustained a slight injury to a finger from a fall. The finger is healing very nicely, we are pleased to report.")

During the 1940s, Sammy Sautin's new Howes Cave Hotel and its adjacent two-story dance hall across from Main Street were the hot spots in which quarry workers let off steam after their shift or on special occasions. Square dances were held every Saturday night from 1940 to 1947, remembered Art Nethaway in a 1983 edition of the Cobleskill *Times-Journal*. Admission was 25¢ for men and 10¢ for ladies, wrote Art, who played violin in a musical combo that called "the squares."

The Howes Cave Hotel drew a boisterous crowd at times. "There were a lot of fistfights at Sautin's," Pangman remembered. "They'd fly right through the front window and into the parking lot."

Boom Times Ahead

Pangman would get together summer nights with the other young boys from the neighborhood and pick the best vantage point on the hillside around Sautin's. "We'd lie on our bellies and watch the fights from the hill above," he said.

Sautin's also had a poolroom, and you could get your haircut there during the day. For 10¢, you could get a cut "that looked like someone put a bowl on your head," Pangman recalled.

Sautin's also let the quarry workers keep a running tab at the bar, at least until payday. As a result, one former resident remembers the housewives of Howes Cave getting much less of a paycheck than could be expected.

A postcard view of the powerhouse at the Howes Cave plant. Built in 1926, excess heat from the plant was released over two cooling ponds. Security guards kept the kids in Howes Cave from swimming in the warm ponds year-round

After the war, business picked up again substantially at the quarry, as well as at its namesake business to the north, Howe Caverns. The tourist attraction employed many dozen local teens each summer; at its peak, the cave drew nearly a quarter-million visitors each year.

Bob Walker, whose dad was a plant foreman at the quarry, worked at the cave as tour guide in 1948, but not for long, a friend recalled. A resident of the hamlet of Howes Cave, Bob took the most convenient, most direct route to the Howe Caverns tourist trail. With only a flashlight, he walked, stooped, and waded each morning through the 1,800 feet of "old" cave that links the quarry with the modern attraction; Bob would then just wait at the end of the underground Lake of Venus for the first tour of the day to pick him up.

That didn't last long, remembered former Howe Caverns board member John Murray Sr. The caverns' management discouraged Bob's underground

Students in the two-room schoolhouse in Howes Cave are pictured here. Margaret May Provost, who was married in the cave to contractor Roger Mallery in 1927, taught there for many years.

Boom Times Ahead

route and, after he took it one or two more times, "that was the end of Bob," said Murray. And although fired from his first job as a teenager, Dr. Robert Walker later enjoyed a distinguished career in science, Murray said.

In 1950, Howes Cave was in its heyday. It was a thriving community with a two-room schoolhouse; two churches (Reformed and Roman Catholic); Maxwell's Chevrolet dealership and garage; Tillison's Luncheonette and grocery store; a boardinghouse; Scottie's Newsroom; Mickle's Grocery Store; the train depot and freight station; and Sautin's Howes Cave Hotel. The post office on Main Street, opened in 1867 in a 12-foot-by-20-foot renovated ice cream parlor, was a busy place, although Howe Caverns accounted for the bulk of its mail. In the mid-1970s, it was documented as the second smallest post office in the United States, then serving 306 families.

That year, the quarry hit its own high-water mark, producing and shipping a record 536,000 bags of cement in a single month, June. Throughout the decade, North American was the county's largest private employer, boasting a workforce of between 200 and 240. Times were good.

And despite a stone quarry, heavy rail and truck traffic, and several kilns and smokestacks, the little community of Howes Cave remained, according to one former resident, "a nice clean village. The houses were all painted and kept well."

Tales of Tite Nippen

Where'd That Name Come From?

The North American plant offered the best-paying blue-collar jobs in Schoharie County. There were 3 shifts keeping the kilns running 24 hours a day, 7 days a week.

In addition to good pay, the benefits were first-rate and there was always plenty of overtime, which "provided big checks and above-average earnings," recalled T.L. Wright of Cobleskill, a ten-year veteran of the plant, 1950–61.

"The jobs at the quarry were hard to get unless you were related to a worker or had someone on the inside to vouch for you," said Wright (1930–2010). Employees had to join the Cement Workers Union Local #65, which awarded jobs through a bidding process and based on seniority.

Shift changes were announced by a loud whistle blast. Employees were provided modest lockers and shower accommodations to clean up after their shift.

"The cement plant was a hot dirty place to work," remembers Wright, "but as the employees said, 'the money was clean.'"

Employees' wages ranged from $1 to $2 per hour in the 1950s, depending on skills and seniority.

The company treated employees well in other respects. Workers were applauded for safe practices and honored for their years of service; they were sent off with well-wishes, cake, and parties when they retired.

Tales of Tite Nippen

Employees' achievements, as well as those of their family members, were chronicled in a company newsletter filled with births, hospital stays, and other "newsy" items.

There were company-sponsored teams for bowling and softball, and company picnics in the summertime. The church pews were filled on Sunday.

With prosperity, there grew in Howes Cave a growing social distinction between the residents on opposing sides of the Cobleskill Creek.

Nearest to the cement plant, to the north, were the newer homes, the Main Street business district, and several company-built homes for North American plant superintendents.

To the south of the creek was Tite Nippen, a collection of older homes built in the late 1800s by the first company to process cement in Howes Cave. Instead of a housing perk for skilled employees, these two-story, two- or four-bedroom homes over the years had become ill-maintained and were now housing the quarry's laborers. The residents there were considered "backwoods" by some.

Tite Nippen was the wrong side of the tracks, or in this case, the creek. Mothers warned their children not to cross the footbridge that connected that part of the community with the rest of Howes Cave.

A snow day, probably mid-1930s, in the "Tite Nippen" section of the Howes Cave hamlet. The quarry is just out of site at the top of the photo.

"The living conditions were not up-to-date or sanitary," posted Bob DeRuvo to an online bulletin board for Cobleskill High School alumni. "But as kids, we didn't know that."

"I think about it now and I don't know how we all survived," said Shirley LaBadia, another former resident of Tite Nippen. "None of us had a whole lot, but I do remember a lot of good times as a child."

Tony Spenello Jr., remembers a cold wind "blowing the curtains flat out" in drafty homes, along with wooden sinks fed by lead pipes.

All the "Howes Cave Kids" went to a two-room school with Margaret Mallery as the teacher. Regardless of what side of the Cobleskill Creek they lived on, they formed a tight, lifelong bond.

"We all grew up and we realize what we had as children was not so bad," Lavina Mulbury and LaBadia wrote to the Schenectady *Gazette* in the fall of 2005. They were replying to a columnist who they felt had depicted life in Tite Nippen as more unsavory than perhaps it was.

"Homes were probably not that much different than other homes in small villages like Howes Cave," they wrote. In the 1940s and '50s, not many homes had running water or bathrooms. "What small village did?" they asked.

Mulbury and LaBadia also took exception to the impression they felt the *Gazette* columnist left on readers and thought he exaggerated the good-time appeal of the local watering hole, the Howes Cave Hotel. "You made it sound like they [the Tite Nippen quarry workers] were a bunch of no-good drunks," they wrote. "These men were a bunch of hard-working men . . . We're sure they did not drink any more than the men on the 'other side' of the tracks."

The origin of the name "Tite Nippen" remains a mystery to this day, perhaps lost to history. Labadia asked a number of locals about Tite Nippen and said that the most logical response she received was that it was a reference to residents' financial straits, as in "money's tight."

DeRuvo heard the name was descriptive of the difficulty entering or leaving the area, as it was a "tight nip in."

"But I'd sooner think it had something to do with booze," he wrote.

Tales of Tite Nippen

Only the foundations of the Tite Nippen houses remain today. They were torn down in 1957 and '58. The area is now overgrown, although some acreage is now part of the Town of Cobleskill's Doc Reilly Community Park, accessed from State Route 7.

The end of Tite Nippen happened to coincide with the end of the rowdy Howes Cave Hotel. Sammy Sautin's popular bar for quarrymen was purchased in 1951 by Alex Bautochka, Sr., who sold it to his son Michael in 1956. A few months later, North American purchased the hotel, barber shop, and the street; the following year they tore it all down. The sign, "Howes Cave Hotel," eventually found its way to the barroom of Boreali's Restaurant on Route 7, proudly displayed by owner Fernando "Fred" Boreali.

As the quarry headed into the decade of the 1960s, the plant was becoming older, less efficient, and more costly to operate than the new modern quarries that were springing up along the Hudson River to the south, notably Marquette, LaFarge, and Blue Circle, to name a few.

A photo looking into the Penn-Dixie Cement quarry at Howes Cave in the mid-1960s. The Cave House can be seen at the far right; in the foreground is a 9,000-square-foot maintenance building; behind that is the huge "crane building," 400 feet long and 120 feet wide.

And America was moving less and less freight by rail. Since the second half of the nineteenth century, the Howes Cave plant had been well served by the railroads and at one point had five separate lines of track to the plant.

Being able to move huge bulk loads of cement by barge, the cement makers along the Hudson had distinct competitive advantages. And as the Delaware and Hudson line that served the North American quarry became less and less dependable (the D & H eventually going bankrupt), the plant moved to a smaller, pre-weighed bagged product that could be more easily shipped by tractor-trailer.

Still, it was more costly to ship cement by truck than by rail, especially over longer distances.

Forest Wollaber Jr., of Cobleskill (1930–2014) claims to have hauled the first tractor-trailer load of bagged cement from the Howes Cave quarry on March 9, 1959. He remembered arriving in the early morning hours between shifts and pulling off the road for a quick nap. Plant Manager Curtis Planck tapped on his windows to wake him, climbed in the cab, and rode with him to the loading station for the initial shipment.

Hard Times

Competition and Environmental Challenges

In 1961, the Howes Cave cement plant was again purchased, this time by Marquette, which sold out less than two years later to another industry giant, the Penn-Dixie Cement Company.

Marquette's first move was to cut costs, reduce its property tax burden and, in general, get rid of what they considered excess. The company is not remembered fondly in Howes Cave.

Like the very first Howes Cave cement company, Marquette's predecessor had built housing for its top employees—six identical two-story stucco homes within walking distance of the quarry. Built in the 1920s, they were offered to department managers and their families for just $8 per month.

Marquette decided to do away with the housing perk, and in June 1961 notices were mailed to each family telling them to get out or to buy the homes. Renters were given a one-year deadline in which to make their decision.

Kiln supervisor Tony Spenello Sr., and his brother Thomas had another idea. They purchased five of the homes for the company's asking price, just $50 each. One-by-one they put them on a large flatbed truck and moved them about a half mile to a new location in July 1962. Each of these "Spenello Drive" homes off Route 7 and Sagendorf Corners Road is still occupied today.

Penn-Dixie expected the Howes Cave quarry to boost the large corporation's overall earnings in 1964, only about a year after they purchased the plant. The previous year, the firm—with 16 plants and 8 distribution centers nationwide—reported a slight decline in earnings, to $3.5 million on sales of more than $50 million.

Purchasing the Howes Cave plant, according to company president Fred Doolittle, would enable Penn-Dixie "to expand its New England markets and serve portions of New York State more effectively." Penn-Dixie employed 136 hourly employees and 26 salaried employees in Howes Cave.

In addition to producing nearly 2 million barrels of cement, there were 596,000 tons of stone and shale produced each year, according to the company's year-end report. In the manufacturing process, the plant used 79,000 tons of coal, 11,600 tons of gypsum, and 5,000 tons of iron ore.

Two years later, more than $500,000 in improvements was announced to make the plant more efficient. In a January 27, 1965, article in the Cobleskill *Times-Journal,* Plant Manager John Weinerth described plans to mechanize the loading system and repair the dust collectors, which were creating a growing concern.

The new loading system included the mechanics to draw cement from any one of the ten huge concrete silos. After loading, the tractor-trailer or rail car would move to one of the two new scales that were installed. After weighing, delivery tickets for billing were then stamped and printed automatically. Additionally, new control consoles located in the scale houses enabled the operators to perform their duties without leaving their inside stations. (For years, the truck scale for smaller loads was directly in front of the company's offices in the Howe family's old Cave House Hotel.)

A better system for collecting the dust that is inherent to the cement-making process was sorely needed. The "dry process" that had improved efficiency in the 1950s was creating more dust than the old "wet process," and a gray film had been slowly accumulating over the homes and businesses in the hamlet of Howes Cave. As production increased, so did the dust.

The 1965 improvement plans described in the newspaper called for the complete repair and overhaul of one of the large kiln's dust collectors. The work consisted of structural repairs and covering the entire kiln's collection unit with stainless steel sheeting. Plans called for the redesign of the dust withdrawal screws and duct work.

And while company officials stressed that the dust pollution was not a problem, employees and quarry neighbors thought differently. (After repairs were complete, a company spokesman proclaimed for a local newspaper headline, "Dust Gone at Howes Cave.")

"Employees had what they called 'cement cars' that they drove to work," remembered Cobleskill's John Pangman (1931–2020). "The dust was so thick that workers would scrape it off their cars with a snow scraper." A separate, clean, car was used for all other transportation.

The most prominent building on the quarry grounds is referred to as the "Crane building." This was 400 feet long, 120 feet wide and about 60 feet high, and once held 2 overhead cranes, capable of moving multi-ton bins of raw materials, including stone, gypsum, iron ore, and clinker to the next processing stage, prior to the final grinding stage.

Unfortunately, asbestos was used in a number of components that kept the plant running and processing stone in kilns that reached 2,700° Fahrenheit. Braking systems for the huge cranes contained asbestos and often needed replacing; kilns were lined with bricks coated with an asbestos-containing cement; burning pipes were lined with asbestos rope; and workers wore asbestos gloves when working with the kilns.

Fifty years ago, many workers were unaware of the dangers of exposure to asbestos dust and fibers and worked without masks or protective gear. They were at risk for developing asbestos-related diseases such as mesothelioma or lung cancer. Some undoubtedly did.

It was no wonder then, that the relatively new Environmental Protection Agency (created in 1970) was beginning to flex its mandate, and there are reports—unconfirmed, but likely—that the Penn-Dixie plant was ordered to make substantial, costly renovations to bring the plant's dust emissions down to acceptable standards, less than a decade after the 1965 improvements.

The U.S. cement industry faced competition from overseas in the 1960s, and Penn-Dixie was no exception. On top of that, there were ongoing labor disputes between the plant's owners and the Cement Workers Union, longtime employee Ken Pangman of Cobleskill recalled.

Richard ("Rick") Wick Jr., a member of the Cobleskill High School class of 1974, had hoped to go to work for Penn-Dixie after graduation. He had the necessary "in" at the plant. His grandfather, Leo ("Ike") Wood, had been a general foreman at the quarry, and his dad worked there as well. The quarry was a short drive from the Wick home, and Rick Jr. worked summers as a teen at another neighborhood business that took advantage of the area's limestone underground—Secret Caverns.

But Rick heard the rumblings among quarry employees and rumors of what was ahead. After graduation, he decided against following his father's footsteps and took a job that offered a more secure future at the State University of New York, Cobleskill.

In 1976, Penn-Dixie shut down most of the operation, putting nearly 140 employees out of work and for all practical purposes ended the century-old history of cement making in Howes Cave. Flinkote Cement of Stamford, Ct., restarted the finishing operation later that year and ran it with a staff of 17 until shutting down in 1986.

Three years later, a tornado tore a portion of the roof off the Cave House Hotel that had been used (and badly remodeled) as cement company offices since the beginning of the twentieth century.

For the next twenty years, the quarry and hamlet of Howes Cave were virtually deserted.

The county's animal shelter moved into the old post office building, and Tillison's Luncheonette on Main Street continued to operate until the death of the owner, George Tillison, in 1993. Although customers were few and far between, you could still get a few groceries, breakfast, and lunch—soup and a sandwich—at Tillison's in the early 1990s. While the sandwich menu was limited, diners had plenty of soups to choose from—whatever Campbell's variety they saw on the grocer's shelf.

Some Historical Firsts

A Chronology of Innovation

While the discovery that finely powdered stone can be mixed and used as a binding agent predates modern history, the processes that turn limestone into cement have steadily evolved to make it better, faster, and less expensive.

Today's cement is so fine that one pound of it contains 150 billion grains, according to the Portland Cement Association. Making it requires an exacting process of some 80 separate and continuous operations, the use of a great deal of heavy machinery and equipment, and huge amounts of heat and energy.

Many of the manufacturing gains made by the industry over the last century have been to reduce energy use and/or to conserve and reuse that heat and energy.

Although not the longest continually operated plant in the U.S., the cement works at Howes Cave (1867–1976) kept pace with these evolving technologies and, in some instances, set the standard for others to follow.

A number of industry firsts took place in Howes Cave between the turn of the century and 1925, when the North American Cement Company took over the operation.

- At the turn of the century, the plant used oil for firing the kiln that burned the stone into clinker for processing. Cornell

University Professor R.G. Carpenter, after visiting the plant, developed a method of burning pulverized coal instead of oil. His patent had applications throughout the industry.
- The plant was the first in the U.S. to use the "wet process" with limestone and clay. This made it easier to grind the mixed raw materials (the slurry) before they enter the kiln.
- It was the first plant to use compressed air to agitate the slurry.
- It was the first plant in the U.S. to be fully equipped with induction motors, which allow operators to mix the slurry at variable speeds, among other uses.
- It was the first plant to use pressure coolers, which compress gas to cool the clinker more quickly as it leaves the kiln. (This is the same process that makes refrigerators and air conditioners work.)

These historical firsts, at varying times, made the Howes Cave cement plant a "state-of-the-art" operation that used the most efficient, most productive and cost-effective manufacturing processes in the industry. Adding to its market appeal, the naturally occurring limestone beds of Howes Cave produced cement that was nearly unmatched for its strength, durability, and uniformity.

Appendix
From Stone to Cement

The manufacture of cement has changed little since the ancient Egyptians mixed lime and gypsum mortar as a binding agent to hold together the pyramids. Over the centuries, particularly in the 1800s, different blends have added to the mixture's bonding strengths and durability, while industrial and technological advances in its manufacturing have simplified its production.

Until about 1905, the only type of cement taken from the Howes Cave beds was "natural cement," made from the naturally occurring mixture of limestone and clay. These were the 20–40-foot-thick limestone deposits found below ground and, of necessity, dug out or mined.

Digging for limestone in upstate New York was not as hazardous as mining coal at any Pennsylvania mine of the same era. Still, a miner's life in Howes Cave in the late 1800s and into the early twentieth century was not an easy one.

A typical mining crew consisted of about a half-dozen men who spent most of the day underground in a chamber of approximately 20 feet in length and width, with ceilings limited to about 7 feet in height, supported at intervals by heavy wooden braces. These rooms were dimly lit. A few candles were placed at strategic rock alcoves, and gas-burning miners' lamps worn on cloth caps provided some light to the working area directly in front of the wearer. The lamps held chunks of carbide, which when mixed with water produces acetylene gas. The gas escapes through a pinpoint nozzle and ignites. The flame casts a reflective light from a chromed disk. It smells like rotten eggs.

Heavy, hand-held pneumatic (air) drills—fed by long lengths of hose—were used to bore holes into the thick limestone walls; explosives were set and charged. Once blasted apart, blocks of stone were loaded onto carts and hauled to the surface by mule.

The miners breathed fumes from the explosive blast and limestone dust. The temperature was a constant 52°, and the mines were constantly damp.

Once back on the surface, the stone was hauled, again by mule, to a kiln for burning. Heating limestone to 2,700+° Fahrenheit creates a chemical reaction in the stone and breaks it down to smaller-sized rocks called "clinker." At this stage, various additives—clay, shale, gypsum, etc.—are added; the clinker is then cooled and taken to the water-powered mill where a heavy Esopus stone ground the material to the finished, fine powder.

Once dry, it was branded, bagged, and shipped.

A view into Cobleskill Stone Products' Howes Cave quarry from the east. The Cave House is on the right; the Crane Building, which houses the stone-crushing unit, is seen in the background. Also shown is the smokestack from a previous quarry owner's rotary kiln, used for burning limestone. It stands more than 100 feet high.

SECTION III

The Cave and Quarry from 1990 On

Howe Caverns from 1990 On

Pushing Beyond the "Mystery Passage" for More Cave

Beginning in the late 1970s, Howe Caverns, Inc. saw its fortunes begin to decline—not significantly at first, but the snowball was rolling downhill. The number of annual visitors, which peaked at nearly a quarter-million a decade or so before, was steadily declining. Travelers were shocked at prices at the pumps following the 1973 energy crisis and gas shortages, and prices never seemed to return to what they once were. Day-travel destinations like Howe Caverns never really recovered. Less than an hour away,

A wide-angle view of the Howes Cave quarry taken on a bright spring day in 2021 from an East Cobleskill vantage point. The Cave House is seen at right; the huge Crane Building, which houses stone-crushing equipment, is at left. The Howe Caverns entrance lodge is just out of frame in the upper-left-hand corner of the photo.

Howe Caverns from 1990 On

John D. Sagendorf, a third generation Sagendorf in the cave corporation, was general manager from 1996 through 2007.

the Capital District (the closest Standard Metropolitan Statistical Area), was not populous enough to draw the crowds the caverns once enjoyed. The caverns' market had always been downstate, New York City, and Long Island, and fed off the other crowds at Cooperstown drawn by the Baseball Hall of Fame.

The caverns' earnings fluctuated and costs increased. Other competition for entertainment grew as more and more Americans just decided to stay close to home. Home entertainment centers, video games, and the beginnings of social media also played a role as America neared and then entered the new century. Educators even started referring to a "nature deficit," as more and more people played indoors, nested, binged their favorite shows, and gamed. Others, a small minority, wanted more adventure in their leisure time, and this gave rise to extreme sports that seemed death-defying to non-participants.

The cost of airfare also dropped, and travel to faraway places became in reach, adding to the decline in attendance at the local attraction.

And, as an attraction that at the time was beyond its 60th year, millions of people had already seen Howe Caverns, and so had their children. And caves don't really change. What could bring them back?

Two long-term managers at the cave left the caverns' employ within a few years of one another. General Manager Rodney Schaefer retired in 1984 and died at his home on the caverns' estate on Christmas Day, 1995. Assistant Manager Harrison Terk was promoted and stayed on as general manager until retiring in '95. Not long after, in 1996, the board of directors brought in a third-generation Sagendorf—John Sagendorf—who came back to the cave after a career with a national contract services provider working in the healthcare and educational fields.

John is the son of Walter Sagendorf, the oldest of John and Mabel's four boys, and was named after his grandfather who died tragically in the cave in April 1930.

The Sagendorf family put down roots in the area in 1802. German immigrants George and Catarina Sagendorf, both born in 1761, built their homestead at what is now the intersection of Sagendorf Corners Road and the appropriately named Caverns Road. The couple had nine children.

John, their great-great-great-grandson, had a natural, lifelong connection to the cave, its corporation, and the cave country. Farm chores in the cave country usually include "stone picking" after a field is plowed. "There were and are many sinkholes on the Sagendorf properties," John wrote in early 2021. "We generally called them 'rock holes' since that was where all the rocks were dumped each year when we went stone picking."

John worked summers and weekends at the caverns from the age of 14 until 1965 when he was 23. He started as a "trailer"—a guide-in-training that listened, learned, and followed behind tour groups. Then he became a tour guide, and was finally moved above ground to the ticket office. (For a number of years, guides were being mocked, as an underclassman might be, if they were called a "trailer.")

"I was fortunate to both have the Sagendorf family connection to the caverns and to live just across the field, so going to work was easy for me," John recalled.

He took over during challenging times. Sagendorf updated the caverns' look with a new logo and billboards, made improvements to the restaurant,

Howe Caverns from 1990 On

and refashioned the attire worn by employees. He took on a few major initiatives as well. One early attempt was to buy out the competition.

At Sagendorf's urging, the Howe Caverns, Inc. board of directors offered to buy the competitor just up the road—Secret Caverns, owned privately since its 1930 opening by the Mallery family, who also happen to own about a fifth of the shares of Howe Caverns stock. The board appointed director Len Berdan of Schoharie as their emissary, and he met cave owner Roger Mallery at his law offices in Cobleskill and made the million-dollar offer.

"We'd never accept an offer to sell," the 89-year-old Mallery said later, in November 2020. The cave remains open today, run by a third-generation family member, Roger's son, known as R.J.

As he concluded "They Bored a Hole in a Hill," the chapter on Howe Caverns in his 1946 book, Clay Perry wrote: ". . . the rather unromantic name of Howe clings to this virtual underground empire that stretches through the hill for one and one-half miles . . . with the possibility of there being other adjacent caverns behind the rock walls of the developed portions. No effort has been made since 1929 to open them and develop them. Enough is enough."

But enough was no longer enough in the new century, and the caverns' management decided more cave than what they already had was needed. They could offer adventure as well.

In mid-August 2005 came the "Adventure Tour," today a $125-per-person, 2 ½-hour tour into a section of the undeveloped cave off the beaten brick paths at the end of the circuitous Winding Way. It's a portion of the cave known since the days of Lester Howe.

Pay-to-cave explorers were (and still are) supplied with coveralls, gloves, boots, knee pads, and a lighted helmet to twist and crawl through Fat Man's Misery, a sandy, two-foot-high belly crawl of about 20 feet. From there the cave opens up—after a hands and knees crawl—into a narrow walking corridor ranging from 5 to 18 feet high.

The Adventure Tour leads about 230 feet to the "Great Rotunda," only seen prior to 2005 by a handful of caverns' employees, cavers, and guests of Lester Howe in the second half of the nineteenth century. The Great Rotunda is a 107-foot-high, silo-shaped natural dome about 10 feet in diameter; Howe, on occasion, would fire skyrockets upward to demonstrate its height.

Caver/karst hydrologist Paul Rubin sits beneath the Great Rotunda, a highlight of the Adventure Tour at Howe Caverns. Photograph by Pete Jones.

Fat Man's Misery is now more appropriately called "The Devil's Gangway," a name given to another tight section of the older part of the cave by Lester Howe. The section of cave leading to the Great Rotunda is now called "The Mystery Passage." The tour package also includes the regular hour-plus cave tour and underground boat ride.

"There are always those who want to see something few others see," said Sagendorf from his Florida home in December 2020. "The saying I often used was, 'There are always those who will pay to sit in the front row!' "

"Some people may be a little apprehensive because it's narrow and they just have the light from your helmets," Sagendorf said in the August 2005 media event. "But it's a comfortable crawl." (Sagendorf tried it himself as a young tour guide in the early 1950s.) "'The crawl was more of a wiggle,'" he said. "'The ceiling was on my back, and the floor was on my belly. It was a pretty rugged journey, not comfortable or suited [prior to clearing the crawlway] to the public."

An Elmira family described the tour on the TripAdvisor Web site:

> There are no lights, except from the helmet on your head. You will be crawling on your hands and knees and getting stuck in the mud. A lot of mud . . . We had so much fun seeing the cave from a unique perspective away from the "touristy" part and getting a sense of the caves that we don't even know about that could be lying beneath our feet.

What Lies Beyond

While this portion of the cave has been known for more than a century, opening it to the public was not an easy task. And the passage, which trends northeast off the main cave, offers other, tantalizing possibilities for more cave beyond. Possibly much, much more.

Caverns' management brought in Paul Rubin, a long-time northeastern caver and karst hydrologist[1] in 2004 to first, open Fat Man's Misery, and then explore beyond known passages and discover options that might open other new sections of the cave. Rubin developed educational programs and materials and trained guides to lead visitors off the clean brick paths into the Adventure Tour. He led the first few tours himself. The first Howe Caverns staff to lead the Adventure Tours included Rubin, Cave Manager Jeff DeGroff, and Mark Tracy, Jessica Dever, and Guy Schivone.

One of the biggest challenges to kicking off the Adventure Tour and opening new cave was clearing mud, silt, and rock from the crawlway entrance (Fat Man's Misery) to the cave beyond. This northeastern route developed over millions of years as a tributary to the main cave waterway, the "River Styx." Most often dry, the passage floods nearly every spring and after intense downpours. It leaves behind mud, silt, and rock.

In the early spring of 2005—well before the tourist season began—efforts got underway to clear the passage. Sagendorf remembers it being a particularly dry year, but spring meant the potential for flooding, so whatever was pulled from the passage had to be moved out of the cave entirely or risk being sent everywhere on the tourist path by a flooded River Styx.

Rubin and Sagendorf got help from about 20 volunteers affiliated with the regional caving club, the Helderberg-Hudson Grotto. They kicked off the dig on March 12, 2005; with a few Howe Caverns employees, they formed a chain gang to dig, fill, and pass 2 ½-gallon pails of mud from the Fat Man's Misery pit. (They were the largest containers that would fit in the crawlway). "It was like doing the work with a teaspoon," Sagendorf told the *Times-Journal* of Cobleskill in August 2005.

The pails, weighing at least 40 pounds each, were hoisted over the railing and dumped into one of at least a dozen 55-gallon drums stocked by the entrance of the tunnel that leads about 250 feet to the elevators to the surface. The heavy mud had to again be taken from the barrels and put into 5-gallon buckets to wheelbarrow through the tunnel and up the elevators. (The intent was to keep the mud from being washed or tracked elsewhere in the cave before it could be removed.)

After the initial dig by volunteers, it took 6 weeks to finish clearing the passage and haul the mud and rock to the surface. In all, 23 tons of earth had been dug and hauled before the first Adventure Tour could be offered. Now picture the efforts made during the 1927–29 efforts to reopen the cave.

With the Adventure Tour now a regular part of the options the cave offers visitors, Rubin began looking at other parts of the cave that could yield new passages, and ultimately new tour routes. His professional background allowed him to see the cave much differently than the average tourist.

"Some of my very best experiences were spending hours alone in the cave after all the tourists left, seeing and puzzling over the geologic tales whispering from the rocks and sediments," said Rubin. "... and of early visitors awed by bright flares in a faraway dome, and of times with my young daughter, Julia, assisting at my side."

Rubin, of Tivoli in Duchess County, considered a number of possibilities. He documented what many northeastern cavers have long thought possible—there is potential to connect McFail's Cave to Howe Caverns

This aerial view of the caverns' estate from the late 1940s–early '50s shows the surface area above the cave, which runs parallel to the east-west road leading to the lodge. The potential for additional cave may exist to the west, just beyond the lower parking lot, and to the northwest beneath the wooded area.

through a "missing link" of underground passage. McFail's Cave, to the northeast of Howe, is the longest cave in the Northeastern United States, at about 7 miles.

Rubin found other options as well, including a tempting, long-overlooked fissure along (and just above) the tourist route, and a sinkhole just off the parking lot.

In the cave country, there is almost always the opportunity for "more cave." To understand where that might be found in Howe Caverns, it is first necessary to finish describing the known section of the cave beyond the Adventure Tour.

Beyond the Great Rotunda is "The Lake of Mystery (on which no man has ever sailed)." It is at this point the Adventure Tour ends and guests retrace their steps.

An underground lake, with few exceptions, is rarely a lake in the conventional sense. No man, or woman, has sailed on the Lake of Mystery for good reason. This underground body of water is about 20 feet long, 5 feet across, and knee deep at best. The cave passage here is between 2 and 3 feet high.

Bob Addis of Scotia, now in his mid-70s, was a Howe Caverns tour guide in the 1960s when he and other guides became the second twentieth-century group to explore beyond the Lake of Mystery. The crew surveyed and mapped this part of the cave in November 1965, and Addis wrote about it for the *Northeastern Caver* in May and July 1969. The article was picked up again in June 2012, and portions are included here to describe the northeastern section of the cave beyond the Lake of Mystery.

The survey team, Addis wrote, ". . . progressed through a crawl, the Lake of Mystery, up over some breakdown, and up a short climb to the right into the Flowstone Rotunda." [Breakdown is just what it sounds like—portions of the caverns' ceiling that have fallen.—Ed.]

The Flowstone Rotunda, Addis continues, "is twelve feet in diameter and at least forty feet high (we had trouble seeing the top of the dome). This dome is completely covered with flowstone in various hues. The Flowstone Rotunda is very decorative and most impressive. A crawl leads off from here, and quickly diminishes in height.

The explorer soon finds himself crawling out on a huge level block of breakdown. When the height is about twelve inches, a crack to the right leads down into Bortell's Lake. Once through the "Crevasse" and into the water, the total height is two feet—one foot of water and one foot or less of air space...

After crossing it [Bortell's Lake] and straight ahead, is the "Rockfall" which is impassable to all courses except the stream. At the T-shaped junction one discovers a stream flowing from Smith Lake on the right to a short stream passage on the left and into the Rockfall.

[*These underground lakes are named for the caverns' employees who first crossed them, Bill Bortell and George Smith, in 1955. Reynolds' River was named for head guide Don Reynolds, who led the explorations.*]

Addis continues:

The far shore of Smith Lake is marked by a strange speleothem—a column about a foot in diameter and three feet high, located in the center of the stream. Beyond this is the attractive Flowstone Gallery with erosional potholes in the streambed.

Most of the remainder of the upstream section is a thin vertical fissure about eight feet high. It either has an hourglass cross-section or else breakdown fills the midsection, forming two levels.

Giovanni's Passage [named for another guide on the earlier trip] is a small, pinched-off fissure, leaving the main passage at chest height and containing a tiny stream. It ends after a few feet.

About five feet above the stream on a mud bank are the initials of the 1955 trip members—Reynolds, Smith, and Giovanni. This marks their furthest explorations...

Going further upstream requires traveling in the nine-foot fissure, split into two levels—the stream and the upper levels. At a four-way junction, the main passage is crossed at ceiling level by another passage, both upper directions being blocked with breakdown.

The exploration continued for at least two or three hundred feet beyond the four-way junction, stopping only for lack of time. The upstream section continues, passes some side passages, and becomes a bit larger due to lack of breakdown.

At this point, time constraints forced the Addis surveying party to stop, and the explorers retraced their steps. It would be another 20 years before Addis would have another chance to explore beyond the Winding Way, and again in 1989 with the author and two other caverns guides for photos for *The Remarkable Howe Caverns Story*.

Rubin described his own efforts, the intriguing possibility of a "Missing Link," and picks up on where Addis left off in an article he wrote for the *Northeastern Caver*, June 2012. The account has been edited slightly to ease understanding.

He wrote:

> Mark Tracy [a caverns' employee] and I pushed to what we believe to be the upstream end of the Reynolds River passage—as far as we could go without dive equipment. It is possible that Bob Addis may also have reached this point previously.
>
> The "end" for us was found after snaking our way along a narrow canyon for a considerable distance. The end was a nicely enlarged elliptical joint—a common, fairly-straight or vertical passage—with a pool at its base. Below [the] water was what appeared to be a conduit, perhaps two to three feet in diameter, extending perpendicular to the joint in both directions. It presented that very inviting look that one would love to have had with a face mask and maybe a [scuba] diver's pony bottle—a small, air supply, typically for emergencies—to take a quick look.
>
> Reynolds River is a wondrous place for two reasons. First, here alone Howe's Cave maintains all of its original glory—wild

and untamed, often fringed with beautiful pristine formations. Yet the real exciting portion—the "treasure"—does not lie upstream. Instead, miles of unknown cave beckon perhaps only tens of feet southwest of the Rockfall.

The Rockfall is nothing more than a large pile of breakdown that obstructs a large passage. Joe Armstrong and I poked around here on another trip while radio-locating it with Brian Pease on the surface. Joe pushed to a point where it appeared that open passage was visible a short distance ahead. Some rock removal would almost certainly be needed before reaching the natural bedrock passage walls beyond.

Rubin ponders the exciting possibility:

Once through, cavers would have a choice of which direction to follow. They could go downstream to the left for a short distance until reaching the short upstream portion of the blocked-by-water-filled passage that rises as the River Styx. Even with dive gear, divers would likely not get far before the passage becomes too tight, narrowing down to about fourteen inches, maximum—the length of the long flashlight Ben Guenther used to carry. [*Read more about Guenther in Section III, Chapter 7.*]

We know this because during a very dry time, I ran a pump and seriously dropped the water level, exposing a formerly submerged room with passage continuing as a high, narrow joint. I went a short distance into it but stopped because I had no second pump for back-up. Clearly though, this narrow fissure that funnels the rise of the main cave stream was probably formed as a bypass around the West Passage collapse. [*The West Passage is seen immediately to the left as visitors exit the elevators. It has been "written off" by explorers. Any passages that lay beyond are considered unreachable. —Ed.*]

Rubin continues:

Once through the Rockfall, a right turn in the upstream direction will take the explorer directly into the "Missing Link"—perhaps minus a few minor collapse areas—two miles straight-line distance to the Southeast Passage of the McFail's Cave system, not to mention another few miles uphill, with "feeder streams" leading to the base of all those rock- and debris-filled sinkholes that folks have been digging in for years. [*The author was one of those folks, trying to find new cave.*]

Rubin enjoys the fantastic possibilities:

From [what is seen in] portions of Howe and McFail's, we can pretty much predict that the route upstream would be pretty "boring" [he jokes]—large passage averaging some thirty to forty feet wide by six to twelve or more feet high with a stream meandering along at its base . . . glistening formations especially along the right wall and, of course, the relict carved initials "LH" in flowstone.

The Rockfall represents a true exploration challenge and is a sure doorway to the Missing Link.

Rubin offered another tantalizing possibility—a nearby "doorway" he thinks may directly connect with the main cave passage beyond the Rockfall and what leads beyond that.

"This from the Silent Chamber, just beyond (but above) the climb down into Fat Man's Misery on the tourist route. Look closely at the scallops—a small, dish-shaped depression—along the wall here, near some of the old carved initials."

Rubin's trained eyes see where water once flowed, eventually plugging the tight fissure with sediment; it served eons ago as an overflow route for the main underground stream course. "There is a fair potential that this conduit curves for several hundred feet to a point just beyond the southwest end of the Rockfall," he said.

Back on the surface, Rubin's other effort was to excavate a sinkhole just north of the employees' parking lot that had for decades been used as a dump. Employees would occasionally find plate metal bumper hangers, road signs, and other "memorabilia" among the construction debris.

Digging and finding existing cave passage at the bottom of the sinkhole dump—named Lester's Door—would bypass the collapsed West Passage and lead, potentially, to that Missing Link that connects Howe Caverns to the longer McFail's Cave. "We dug into a wide . . . joint filled with glacial till," said Rubin. The effort was dropped after the winter season.

Rubin had a helper at times—his daughter Julia, at the time 10 years old. "She loved to write the notes in my field notebook," he said. "She even

In 2004–05, Howe Caverns undertook serious efforts to find new sections of the cave to open to the public in hopes of attracting new visitors. Karst hydrologist Paul Rubin of Tivoli led the efforts, which included digging open a sinkhole filled with construction debris near the employees' parking lot. Silt and mud clogged this promising passage, named "Lester's Door."

did a little digging with me in the cave after hours to help check out a promising lead. She loved it."

One evening, Rubin took Julia into the cave after hours to see flooding water rush from the Winding Way, just past the elevator entrance. "As she clutched my data-riddled field notebook near the railing, it slipped from her hands and fell into the raging water below."

Efforts to find it downstream after the floodwaters subsided days later were not successful. "She was, no doubt, really concerned about this. I told her it was all right; we would retake the data points—and we did, with her diligently recording sinkhole and other coordinates.

"It's one of those things where the memory just brings a smile to your face," Rubin shared.

Rubin's efforts below and above ground to expand the cave never reached fruition. There are no previously unseen parts of the cave available as a tour offering today, although other options have been developed to go beyond the lit bright paths and/or see the cave in a different light.

Things were happening beyond the far end of the cave, though, around what was once the natural entrance and the Howe family's old stone Cave House Hotel. These were beyond the control of the Howe Caverns corporation—the mineral rights to the old Penn-Dixie cement quarry had been leased, and a local rock products company was planning to restart operations there after a 30-year lapse of activity.

1. The science of karst hydrology studies the movement of water underground through areas of karst topography, or "cave country." It is especially important to the understanding and protection of water supplies for both public and private use.

Rebirth of the Howes Cave Quarry

Bold Plans Connect Present with Historic Past

The century-old cement-making industry in Howes Cave ended abruptly in 1976, when then-owners Penn-Dixie Cement Company[1] shut down operations, throwing 137 employees out of work (see Section II).

The infrastructure, steel, and equipment that could be salvaged were sold. It was not an easy task. George Steeves of Gilboa in southern Schoharie County was part of a crew that took apart the huge kilns for salvage. He remembers the dimensions of one, weighing hundreds of tons. The kiln measured 278 feet long and 12 feet in diameter; it was lined with 3-inch-thick steel and 6-inch fire-resistant brick.

Schoharie County officials were left to ponder what to do with an abandoned 32-acre hole in the hillside and another 300+ acres unused on the surrounding hillsides still owned by a Penn-Dixie affiliate. Left behind was "a disheveled mixture of broken buildings, equipment and waste," said the author of "Reincarnation of a Quarry" in the May 2004 *Pit and Quarry* trade publication.

Then, in 1984, the county's Industrial Development Agency (IDA) began picking up quarry property in bits and pieces in lieu of back taxes, hoping in some way to restart even a small part of the local economy. The results, at first, were mixed.

The IDA invested in some rehabilitative efforts, clearing and renovating a former stock house (once used for raw materials) and the machine shop and crane building, but the efforts to place tenants there barely broke even. A recycling plant took up space in the huge crane building for a few years, and a beverage distributor rented space, and has since 1992, in a 9,000-square-foot maintenance building. An eccentric group of young entrepreneurs referred to as the "Bungee Boys" manufactured bungee cords in an old Quonset hut.

"It was a dust-covered, depressing example of industrial archaeology," said one community planner.

There wasn't much left in the tiny hamlet of Howes Cave, either. Quarry buildings, scattered throughout the community on both sides of the railroad tracks, were abandoned, in disrepair, and coated with cement dust. The longest holdout on Main Street, Tillison's Luncheonette, closed in 1993 after the death of owner George Tillison. Even the tiny Howes Cave Post Office left the hamlet's Main Street for new offices along Route 7. In 1989, a tornado—quite rare in these parts—tore apart the roof of the Cave House, exacerbating its deteriorating condition.

Lester Howe's former hotel, the Cave House, long-abandoned in this photograph from about 2001. The original entrance to the cave is to the left, below ground level.

Rebirth of the Howes Cave Quarry

Clemens McGiver, one of ten children born to actor John McGiver and his wife Ruth, of remote West Fulton, Schoharie County, has been captivated by the quarry all his life. "The high quarry walls, the monumental scale of the buildings always fascinates me."

"It was the matter of curiosity people always feel about ruins," McGiver said years later in a February 2003 story in *Rock Products* trade magazine. "And when the sunlight fell on [the quarry walls], you could see different formations and strata from eons of time." The industry trade publications would give considerable press to the quarry's restart and its unique mixture of mining, history, and education.

In 1983, working toward his master's degree in architecture at Rensselaer Polytechnic Institute in Troy, New York, McGiver proposed as his thesis, "Adaptive Reuse of the Howes Cave Quarry." That work, resulting in a degree ten years later, became a blueprint for the quarry's multifaceted resurrection ("I'm a bit of a prognosticator," admitted McGiver.)

When the IDA was presented with McGiver's early trove of documents, maps, photos, and models, he was hired as a consultant, given office space in a dusty Quonset hut, and from 1985 to 1989 worked to develop prospects for the abandoned site, or in other words, its "Adaptive Reuse."

The "inherent beauty" in the quarry and the "unnatural landscape" left behind by years of industrial abuse, McGiver felt, offered a lot to work with. "I wanted to do something I could be proud of."

Upstate New York is home to plenty of cavers; when word spread that McGiver had the keys to the original Howe's Cave entrance and portions of the cave long abandoned, he had several offers to help. And McGiver knew the caving community could help him understand the underground mine, caves, and unique surrounding topography in new ways. He felt an instant rapport with Ben Guenther of Richmondville, a tall, thin, 60-something caver with an easy smile and demeanor—and a remarkable mind.

He and Guenther became lifelong friends. "We spent two years together—slowly, creeping along the mines, discovering fascinating artifacts."

It was Clemens McGiver's architectural thesis for the "adaptive reuse" of the Howes Cave quarry and abandoned Cave House that led to ongoing revitalization efforts there. He is pictured at a fireplace inside the Cave House, uncovered during early renovations.

Guenther was brought on as a second IDA consultant, as caretaker. He is credited among other things with mapping much of the 150-year-old mine that underlies the quarry, finding unique rock formations there as well as mule tracks from a century ago, he said, that looked as if they were "made yesterday."

McGiver—described as "a dreamy-eyed architect" by *Rock Products* magazine—immediately recognized the historic asset that was the long-abandoned Cave House Hotel. The hotel once welcomed nineteenth-century cave visitors and later served as cement quarry offices. "It's a Romanesque revival building, with not a lot of ornate detailing," McGiver told *Rock Products*. The cut-stone, early-Victorian hotel, built 1872–73, is "absolutely gorgeous." (See Chapter 7 in Section III.)

Unknown to McGiver and Guenther, others had hopes for the quarry's rebirth as well.

Rebirth of the Howes Cave Quarry

Emil Galasso, the second-generation president of family-owned Cobleskill Stone Products (CSP), had interests in the old quarry for several years, and those interests grew more serious in early 2001.

At the time, CSP was also planning to consolidate its operations to Howes Cave from its older quarry in Cobleskill, about ten miles to the east. New stone reserves would be needed to take the company into the next several decades. The 145-acre Cobleskill quarry is now no longer in use, but CSP still holds its necessary permit from the state Department of Environmental Conservation.

Founded in 1954 by Emil F. Galasso (1916–2010) and his brothers, CSP produces asphalt and crushed stone for paving and does contract work for the New York State Department of Transportation, as well as local municipalities. In addition to quarries in Cobleskill, Schoharie, and Hancock, Delaware County, the company has 11 asphalt plants and multiple sand and gravel pits, as well as affiliated companies throughout the state.

In April 2000, Galasso met with IDA officials to explain his company's plans to restart the quarry, not for making cement, but for making crushed stone products—aggregate. That meant blasting stone from the high quarry walls, crushing it, sorting by size, and washing it for its numerous uses around the region. It's a big, occasionally noisy process.

McGiver and Guenther were at that meeting, listening to Galasso's plans. McGiver remembers, "Ben's face lost all color" before further hearing Galasso's plans for the Cave House as a mining museum and visitor's center that could offer tours of his working quarry. The project, Galasso explained, would be a "natural extension of the company's commitment to the industry and the community."

Although not without detractors, CSP is generally considered as a "good neighbor," well-regarded locally for its community service and charitable largess. Tours of the company's Cobleskill quarry had been offered to educators in the past. Galasso speaks of the tours proudly.

Talking after the IDA meeting, Galasso, McGiver, and Guenther recognized what each brought to the table. The potential was huge. "Suddenly we were working for Emil," said McGiver, who nearly two decades later remains one of the site's most ardent supporters and president of the Cave House Museum's board of directors.

Restarting the quarry would be the economic engine to make the entire project financially feasible, and Galasso was, first, a businessman. He planned a $7.4 million investment to rehabilitate quarry infrastructure and purchase the needed equipment.

"I want people to always remember this is a mine," said Galasso during the early planning stages. "Without limestone," he noted, "people wouldn't enjoy such common products as paint, paper, toothpaste, ink, antacids, and important roads, bridges, and beautiful limestone buildings. We want the community to come and use it [the museum and quarry] and see what the industry can do."

The vision of Cobleskill Stone Products, McGiver, and Guenther was shared by the area's economic leaders, as well as many in the industry, including Caterpillar, Southworth-Milton, and the New York State Construction Materials Association. Former Penn-Dixie employees, professional and amateur geologists, cavers, former residents of Tite Nippen and the Howes Cave hamlet, and others followed developments closely and were eager to see what was coming next.

Alicia Terry, then director of the county's planning and development agency, called it "an exciting project that combines several industries to economically revive a distressed community, create jobs, and provide stewardship for an important geologic area."

Negotiations began with Callanan Industries of Albany—formerly Penn-Dixie—and the Schoharie County IDA to consolidate the old quarry property and adjoining parcels in the Howes Cave hamlet. "I had done several split acquisitions with [Callanan] as well as other joint ventures with them in the 1990s," Galasso said. When he asked about leasing the Howes Cave quarry, Callanan agreed, knowing the abandoned property could again be put to good use.

When the deal was complete, Cobleskill Stone Products had leased approximately 350 acres of the quarry from Callanan for the mineral rights, and purchased 40 acres from the county IDA for about $200,000. (He made the final payment to the IDA in 2012.)

Galasso and his geologists projected the quarry had 50 years of limestone reserves for the company, at a production rate as much as one-half million to one million tons per year of crushed stone for use in concrete, blacktop, foundations, sidewalks, and other construction needs.

"We took an old industrial site that had many environmental issues and cleaned those up," said Galasso when asked in December 2020 of what he was most proud. In the former crane and mill buildings, they installed a state-of-the-art quarry operation.

As part of the deal and at Galasso's request, Callanan donated approximately 20 acres of the site that included the dilapidated Cave House Hotel and about a quarter-mile of old Howe's Cave that had not been seen by visitors in more than a century. An additional 40 acres was also set aside to create a community park along the Cobleskill Creek near what was once Tite Nippen.

Today, the Howes Cave quarry is again active, although at not nearly the levels it was in the 1950s and 1960s. Quarry Supervisor Sam Galasso, Emil's grandson, and co-supervisor Pete Pierce said the plant employs 20.

After two years of improvements that included extensive repair and replacement of the corrugated steel sides of the huge—400 feet long by 120 feet wide—crane building, the quarry went online again in March 2007. Cobleskill Stone Products boasts that the Howes Cave quarry uses the latest extracting, crushing, washing, and distribution technology as it produces crushed stone and stone products for regional sales.

By enclosing the crushing operation in the former crane building, dust is greatly reduced and production runs from early March through November. The mill building, just to the west, and the crane building also shelter the plant's equipment from the elements, said Sam Galasso.

The quarrying process starts with removing the "overburden," or vegetative material, that covers the surface of the stone to be exploded in a tightly controlled blast from the hillside.

CSP blasts about once a week during the summer construction season to meet demand, or in industry parlance, "keep stone on the ground." The blasts, which still can rattle the ground around the quarry, typically knock 12,000 to 15,000 tons of limestone from the quarry face, but varies by the amount of stone needed.

Today's quarry blasts are nothing like they were in the early days of the cement quarry, which could break windows miles away. Those blasts, always after business hours, could rumble the ground at the Howe Caverns picnic area even in the mid-1970s, the author can attest.

Modern blasting techniques use tools that were not available when explosives blasted through and destroyed portions of old Howe's Cave in the early 1900s. Environmental concerns increased dramatically since then, and all mine blasting in New York must comply with stringent limits on the intensity of the shock—vibrations—they release, said independent NYS-licensed geologist Paul Griggs, a trustee of the Cave House Museum. These limits, he said, prevent damage to even the weakest building materials.

CSP and Callanan Industries donated the land containing the Cave House Museum and the Howe-Barytes cave system to the nonprofit museum at the quarry's entrance. As part of their mining permit, CSP has agreed to avoid mining in those areas. The tourist entrances to the two commercial caves, Howe and Secret caverns, are more than a mile away from the quarry, and at those distances, the experts say, can't be impacted by the blasting, which is designed to protect much closer structures.

CSP also agreed to blasting limits that are much lower than the industry standard to further protect the caves and the bat populations that hibernate in them, according to Griggs. Compliance with the state's vibration limit is required of all mining permits.

"CSP operates the Howes Cave quarry in an environmentally conscious and friendly manner to protect the caves and the surrounding community from significant impacts," said Griggs.

Griggs said it was research by the United States Bureau of Mines that resulted in rules proposed in 1980 to limit the strength of the blast and resulting ground vibrations. These were incorporated into all mining permits in New York by the state Department of Environmental Conservation.

Further, all blasts in the Howes Cave quarry—and elsewhere in the state—are monitored by a third party to ensure compliance.

The blast consists of explosives (primarily ammonium nitrate fuel oil and related emulsions) loaded into a series of holes bored into the limestone. In the old days, all the explosives were detonated at the same time. Today, highly accurate delays detonate explosives just a few milliseconds apart. This also greatly reduces the amount of vibration created while still effectively breaking up the rock, according to Griggs.

As of December 2020, most of the stone has been taken from the west face (Cobleskill side) of the old quarry—more commonly called in the industry "the pit"—which is nearest the mill and crane buildings.

After the blast, the "shot"—collapsed stone rubble—is loaded and hauled by truck to the crushing unit and fed into its powerful jaw. At peak capability, the company can crush 750 tons of stone per hour.

Once through the crusher jaw, stone is taken by conveyor to the primary screen, also referred to as the "scalping" screen. At that point, operators have the option of producing stone fill, which is four-plus inches in size, or return it to the crusher to further reduce the size. At the scalping screen, the company also produces "crusher run," a much-used blend of smaller crushed stone and stone dust.

The crusher run stone product falls completely through the screen and onto a conveyor belt. This material is generally used as a subbase for roads, driveways, foundations, and similar projects. Once the material passes the scalping screen, it goes into the secondary crusher where the crushing and screening cycle can be repeated until it is ready to sort into the desired size. There are as many as six separate screens, creating six different sizes of stone.

The material that falls through the fine screens is dust. The dust is sent to a classifying unit that can further reduce the material to a sand-like granule, used for very precise mixes of asphalt. Depending on the final product, the stone may be washed at this point. If not washed, this exceptionally fine stone dust is commonly known as "ag lime" and is often used as a fertilizer in agriculture. The heavier cuts of this unwashed stone, still quite small, are stockpiled as "Buell Dust"—named for the classifying unit manufacturer—an essential ingredient in many mix designs for blacktop.

The Howes Cave quarry provides one-half million to one million tons per year of crushed stone for use in concrete, blacktop, foundations, sidewalks, and other construction needs, according to geologist Griggs, who is also the educational coordinator at the quarry's Cave House Museum. Every year each person in this country uses nine tons of construction materials, mainly aggregate and cement, he said.

Once again, stone from the Howes Cave is being used in important projects around Cobleskill Stone's sales region, which includes ten New York counties and parts of Pennsylvania.

Sam Galasso said in December 2020 that nearly twelve and one-half miles of Route 30A is being rehabilitated from the Schoharie County line to Van Epps Road in Fultonville, Montgomery County. The year-long project started in May 2020 and includes Cobleskill Stone paving, making drainage improvements, and recycling some construction materials.

Nearly 21,000 tons of Howes Cave stone was trucked to CSP's Oneonta asphalt plant to be converted to blacktop and used to resurface both eastbound and westbound lanes and ramps of I-88 exit 13 to Exit 15 in Oneonta.

"We're more than capable of keeping up with demand from the surrounding communities and for our own internal use," said the fourth-generation Galasso. "If market demand remains steady," he said, "we're exceeding expectations."

Cobleskill Stone is already planning for the years ahead by burrowing a tunnel for trucks and other heavy equipment under Sagendorf Corners

---

[1] In 1967, the New York City–based, publicly traded Penn-Dixie corporation purchased the family-owned Callanan Road Improvement Company, based in South Bethlehem, Albany County, founded in 1883. Fifteen years later, the name was changed to Callanan Industries, Inc. Callanan changed hands again in 1985 when Cement Roadstone Holdings (CRH) of Dublin, Ireland, purchased the company through its U.S. subsidiary, Oldcastle, Inc.

During the 1950s the Callanan family company was a major contributor to the construction of the New York State Thruway, supplying over 1.5 million tons of materials for the building of 67 miles of mainline, 31 miles of access roads, 6 interchanges, and numerous bridges. In the 1960s, the Callanan South Bethlehem quarry supplied more than 2.5 million tons of stone for the construction of the Empire State Plaza, which at the time was the largest construction project in the world.

Rebirth of the Howes Cave Quarry

Road to quarry the limestone to the east (Central Bridge hamlet side). "That's where most of the reserves will be," Sam Galasso said.

The tunnel, 150 feet long and 20 feet wide, took about a year to complete, finishing in 2018. It is not currently being used (December 2020). While some stone has been taken from the eastern side of the quarry, Galasso said, there remains good stone to the west.

Left: Crushed stone is loaded for weighing in the Howes Cave quarry. Below: Stone from "the shot" is dumped into the crushing unit for processing.

SOLD!

A Family Says Goodbye

It came as a shock to most Schoharie County residents, to the northeastern caving community, and to hundreds of former employees who have a special place in their heart for the cave when, on April 19, 2007, Howe Caverns, Inc. was no more.

Shareholders in the venerable 80-year-old closed-stock institution voted to sell the shares that many of them or their families had purchased for $100 decades ago or had held from the beginning in 1927. After some initial concerns for what would become of the cave, the shareholders' vote was unanimous.

Many shareholders believed the heyday of Howe Caverns had passed; the number of visitors had been declining steadily since about 1980. At its peak, the cave attracted nearly 250,000 visitors each year, the great majority during the summer months—at times 3,000 a day. When the sale was announced, the yearly number of visitors was down by nearly 100,000.

The Sagendorf family had played a huge role in 1927-29 building the corporation that brought Howe Caverns back from the brink of obscurity, possibly even saving it from destruction at the hands of an expanding cement quarry. Eighty years later, it was a Sagendorf that stepped forward to break up that corporation, intending to preserve the cave for future generations. Ironically, after other offers were considered, the property landed in the hands of a company associated with the new owner of the Howes Cave quarry.

SOLD!

"It was my offer to purchase the business that started that ball rolling," said John Sagendorf, general manager of the cave from 1996 through the 2007 sale. The unsolicited offer took members of the board by surprise in an April 2005 meeting. At least one member was even angry.

Although the number of visitors was way down, directors had never considered selling the property previously. There were grumblings from some stockholders about the drop in earnings, but there were no calls for selling the property outright. No one even really knew what a share of Howe Caverns stock was worth.

"I could clearly see there was not the likelihood of survival as a privately-held stock company," Sagendorf said. "The company had to be too conservative and could not be nimble enough to make the necessary changes for long-term success and growth."

In its statement to shareholders, the board noted that it had "examined the cost to revitalize and improve the facility to modern standards," but concluded those costs were beyond the corporation's ability to handle new debt.

As a closed-stock corporation, shares of Howe Caverns, Inc. had always been difficult to come by, a good indication the cave had been a solid investment for many years. The corporation's board of directors provided fairly consistent dividends to its shareholders and paid "quite well" during the years of highest traffic, several shareholders said.

Cobleskill attorney Roger H. Mallery, who said his family owned 5 percent of the cavern's corporation from the very beginning, was pleased with the return on investment. "It paid pretty well," he said. The Mallery family owns Secret Caverns.

The Howe Caverns, Inc. board, acting with due diligence, hired an Albany firm, the Schwartz Heslin Group, Inc., to evaluate the offer and to solicit others. Seven were received. Shareholders entertained a slightly higher bid, but ultimately rejected the offer from Dongbu Tours, a South Korean company hoping to develop a resort in nearby Sharon Springs. Instead of a one-time buyout, the company offered a series of payments, which shareholders declined. The offer from Sagendorf was also heavily mortgaged.

Another bidder among the finalists put together a $5.2 million offer that would have turned the cave and its picturesque estate into a state park. The Northeastern Cave Conservancy, Inc. worked with officials at the New York State Department of Parks, Recreation, and Historic Preservation during the final months of the Pataki Administration. While Howe Caverns directors were interested in the offer, it failed because of a lack of commitment on the state's part within sale deadlines.

The seven-member board of directors, said member Len Berdan of Schoharie, had been consumed for years with the cave's financial condition. Having a large number of shareholders expecting to profit from a declining revenue source was an "outdated business model and unwieldy entity," according to Berdan, "that had to be sold." Berdan was appointed by the board to coordinate the sale.

Like many with close ties to the cave, Berdan worked there summers during high school and college. His family was among the original investors who picked up $100 shares.

Schoharie attorney Charles M. Wright (1933–2018) and Emil J. Galasso, president of Cobleskill Stone Products, paid $3.7 million to purchase 14,774 shares of the common stock from 220 shareholders. There were some grumblings and concerns, which were assuaged by the Wright-Galasso, Inc. presentation to the board. Each share brought $228.72. (Technically, it was a "merger." Wright and Galasso held 76 shares at the time.)

"I think John Murray [Sr., of Cobleskill] and I were the last holdouts," said former caverns' manager Harrison Terk in a December 2020 phone interview. "We talked it through. Wright-Galasso was a local company, knew the importance of the cave and offered assurances they had good plans in mind."

"It was a complicated sale," said Berdan, "set in place by a series of events, unexpected happenings and strong personalities. It had all the twists and turns of a Victorian novel."

At the April 2007 shareholders meeting, Wright-Galasso Inc. told shareholders they planned to immediately invest $2 million to improve cave lighting, renovate the gift shop, and repair the winding road that leads up the quarter-mile hill to the entrance lodge overlooking the valley.

SOLD!

"Charlie and I have the ability to bring new blood into Howe Caverns," Galasso later told the *Times-Union* newspaper of Albany. "It was at a point where it needed that."

"Charlie loved the cave and held stock," said Galasso. The two men had been friends since Wright had worked for Galasso's father, Emil F., while going to college. "When it [the cave] was falling on hard times, he asked me to partner up with him to buy it." It is a fifty-fifty partnership still in effect with Wright's heirs in 2021.

In addition to the financial commitments, there were other reasons the board decided on the Wright-Galasso proposal. These were included in the terms of the sale. Two key points the new owners agreed to:

- Keeping the caverns open to the public for a period of at least ten years, and
- Employing the corporation's existing full-time permanent workforce after the merger for a period of one year and/or pay severance to those let go within that year.

According to Berdan, the sale couldn't have come at a better time for the corporation's shareholders. Later that year, the nation's housing bubble burst and real estate securities failures challenged financial markets and sparked the "Great Recession" and financial crisis of 2007–08.

"In the interest of the shareholders, the board selected the best [cash] offer. That brought to an end a marvelous, more than eighty-year connection between the caverns and the Sagendorf family," said John Sagendorf, now retired and living in Florida. "I am clearly very proud of the family history connected to Howe Caverns. I was sad to see the business sold and leave the family."

There's still a Sagendorf working at Howe Caverns, and there are still Sagendorfs on the 1802 homestead at Sagendorf Corners Road in the cave country. Sisters Nancy and Mary Sagendorf own the property, and Mary and her husband, Chris Guldner, make the 25-plus-acre property their home. The couple also operates Cobleskill's historic Bull's Head Inn restaurant and pub.

Before his tragic death in the cave in 1930, first corporate secretary John Sagendorf and his wife Mabel had four sons. Those four sons—Walter, Allan, Victor, Willard—had nine children. The majority worked at the cave at some time in their lives.

Willard's daughter Nancy has worked at the cave weekends and summers for 48 years, and continues now in 2020 after retiring as science teacher with the Schenectady City School District. The sale "came with mixed emotions," she said. "How well the cave did financially was always tempered by emotions and pride.

"We wanted the caverns to stay open to fulfill our grandparents' vision of sharing the cave and all its wonders with others," Nancy Sagendorf said. "The impact on Schoharie County and all those who have visited through the years has been and continues to be immeasurable."

Nancy also helps preserve the family's legacy as a board member of The Cave House Museum of Mining and Geology, described in a following chapter.

4

A Word about White-Nose Syndrome

Howe's Cave "Ground Zero" for Deadly Fungus

Some visitors to Howe Caverns and Secret Caverns have an unnatural fear of bats, mostly based on superstition and old wives' tales. Many breathe a sigh of relief when the tour guide explains that Howe Caverns' small colony of bats avoids the well-lit portions of the cave. Once estimated at about 500, these bats prefer the undeveloped section of the cave beyond the Lake of Venus, close to the old entrance, so that at dusk they can leave the cave in search of food.

Bats are one of nature's most misunderstood creatures. Though often feared and loathed as sinister creatures of the night, bats are vital to the health of the environment and economy, according to the U.S. Fish and Wildlife Service.

A single little brown bat can eat up to 500 mosquito-sized insects in an hour, giving them an essential role in pest control. They are also an important link in pollinating plants and dispersing seeds. While many bats eat insects, others feed on nectar and provide critical pollination for a variety of plants like peaches, cloves, bananas, and agaves.

They started dying off in Howe Caverns in the fall of 2005 from a once-rare fungus that infects the skin of the muzzle, ears, and wings of hibernating bats. Believed to originate in Eurasia, Howe Caverns became ground zero for white-nose syndrome (WNS) when its symptoms—not yet diagnosed—were first seen in September 2005. The syndrome takes its name from the appearance of the fungus, which looks like a white, powdery substance.

The earliest hint of a deadly invasive disease among the bats in Howe Caverns was noted by Paul Rubin, a karst hydrologist working with the cave's management to expand the cave tours. A number of dead bats were found both upstream and downstream from the cave's large ventilating fan, some with advanced decay including white muzzles.

By February 2006, Howe Caverns' bat population had been reduced by more than 90 percent, according to Rubin. That month, Ben Guenther, the educational director at the Cave House Museum, found numerous dead, dying, and

Carl Mehling, a senior museum specialist with the American Museum of Natural History in New York City, pauses to take a photo of the bats in the lower section of Howe's Cave in January 2005. At top is an Eastern small-footed bat, Myotis liebii. The devastating white-nose syndrome was first noticed in Howe Caverns in late 2005, about three years before it spread nationwide and in Canada. Photograph courtesy of Kevin Berner.

A Word about White-Nose Syndrome

erratically flying bats woken from hibernation in the lower section of the cave and in an intersecting mine tunnel. He was able to describe the deadly symptoms—waking during hibernation periods, erratic flight behavior, and atypical roosting behavior.

Guenther emailed Kevin Berner, a State University of New York wildlife studies professor whose students had helped with the bat counts. Guenther accurately predicted what was soon to unfold—"Regarding the bats, I believe we have more of a problem than we might imagine."

Since those first observations, WNS has spread throughout upstate New York, northwestward through southern Ontario, Canada, northeastward to Nova Scotia, southward to Missouri and Arkansas, and westward through northern Texas, according to the Fish and Wildlife Service.

The number of bats killed by WNS is unknown, according to the WNS Response Team, a national group of biologists, researchers, land managers, and bat lovers across North America united in the fight against white-nose syndrome. The group represents more than 100 federal, state, and provincial agencies, tribes, universities, and nongovernmental groups.

"Bats were so common in some areas before white-nose syndrome arrived that they simply weren't counted," acknowledged the Response Team. "A group of experts met in 2011 and estimated that 5.7 to 6.7 million bats had died up to that point."

Several mitigation measures are in place across upstate New York's cave country and elsewhere. Most caves are closed during the hibernation season. Information is available online at many sites, including that of the WNS Response Team and National Speleological Society.

5

New Owners, New Outlook

Adventure, an Animatron, Naked Tours, and More

Since the 2007 purchase, Wright-Galasso, Inc. has made a number of changes at the cave intending to draw visitors and encourage a younger, perhaps more enthusiastic, crowd. "It's not your grandfather's Howe Caverns anymore," said Len Berdan, a former caverns board member involved in the sale.

The cave attraction's slogan is now "Keeping Adventure and Exploration Alive."

It is no criticism to suggest that Howe Caverns, Inc. built a conservative brand, based on the cave's unique appeal as a historic, long-standing natural attraction offering families a few hours of "fun and education." The Howe Caverns corporation's twenty-first-century efforts to boost traffic were built around their main asset—they added wild cave tours and tours by torchlight.

Since the sale, Wright-Galasso is presenting a more robust, adventure-park-themed destination with high-tech innovations, a rope course with 26-foot climbing wall and 4-tower zip line, and an H2OGo ball, which (until closed) was a giant water-filled ball that rolled down the entrance hill with a patron inside. There were even more grandiose plans to develop a $450 million casino resort and second indoor water park at the Howe Caverns estate (more on those in the next chapter).

Below ground, Wright-Galasso management expanded the variety of cave tours offered. In addition to the traditional hour-and-fifteen-minute

New Owners, New Outlook

trip, the Lantern Tour after hours on Friday and Saturday nights, and the Adventure Tour, they now offer a "Family Flashlight Tour" on Sunday evenings, a more leisurely "Photo Tour" the third Wednesday of each month, and periodic private tours.

A relatively new offering is the "Signature Rock" tour, a 1,400-foot addition in the Adventure Tour theme. It provides lights and other accoutrements to take explorers over the flowstone dam at the end of the Lake of Venus into another undeveloped part of the cave that heads toward the original entrance. Following the streambed, they pass through historic "Music Hall" and "Congress Hall," to Signature Rock, where Lester Howe and other early visitors carved their names or initials.

Several of the new tour options at the cave attempt to replicate how visitors saw the cave prior to its modern development. One of the newest versions takes visitors into undeveloped cave beyond the Lake of Venus to "Signature Rock," shown in this photo from the early 1900s.

And then there's the Naked Tour, which is exactly what it sounds like.

As owners of a popular natural attraction, it's reassuring that Howe Caverns' managers believe "Natural is Beautiful." Still, it raised a lot of eyebrows—and generated a huge amount of publicity—when it was decided to celebrate International Nudity Day on July 14, 2018.

The public was invited to "Take a Leisurely Naked Stroll by Lantern Light" after regular business hours, while also celebrating body positivity in the 52° cave. For the $65 ticket, guests received a souvenir robe and complimentary alcoholic drink.

The tours have sold out—about 300 persons—each year since, although the tour date has changed. It was last billed as a "Harvest Moon" event held in mid-September. Yes, "moon" was intentional; "Celebrate Your Full Moon," read the promotion. It also became more expensive, $130, but included a three-course meal in the caverns' restaurant.

No cameras or phones—a good policy in cases like this—has been strictly enforced.

Many will be glad that the author cannot offer any firsthand account of what the naked tour is like, so let's rely on the work of veteran outdoors writer/author David Figura. He writes for the NYup.com website and was on that first tour in 2018.

Figura, an award-winning newspaper journalist, writes for *The Post-Standard* newspaper and syracuse.com in Syracuse, New York. He is the author of the 2014 book, *So What Are the Guys Doing?* from Divine Phoenix Publishers.

He wrote the below account for *Underground Empires*:

> I was a bit nervous about the thought of having to take off all my clothes in front of people while at work.
>
> But there I was in the lodge at Howe Caverns in a room packed with smiling, laughing folks in bathrobes, many excited and in some cases as apprehensive as I was about being at the first-ever "Naked in a Cave" event.
>
> Weeks before at a newspaper staff meeting my editor asked for a volunteer to cover the event. The room got silent.

New Owners, New Outlook

I'm not sure what possessed me, but I suddenly blurted out: "I'll do it. It would be something different. I'm not ashamed of my body. It'll be fun."

As the outdoors writer for NYup.com and *The Post-Standard* newspaper, I'm usually covering topics ranging from birding to bear hunting. I had never been to Howe Caverns and figured it was finally time to check things out there.

In the days leading up to the event, coworkers, family and friends were all kidding me. I started having second thoughts about being in my birthday suit in front of what turned out to be about 280 folks from across the state and beyond.

"You can't drop out now," my wife said. "I've told everyone you're going to do it."

Howe Caverns officials had laid down strict rules beforehand for the media. No cameras or cell phones were allowed. Keeping with the emphasis on confidentiality, I only asked for people's first names, their age and where they were from when interviewing them.

On the day of the event, I drove onto the Howe Caverns grounds and stopped first at the small motel there. Brenda, a sixty-five-year-old woman from Buffalo, was among the folks partying in front of their rooms and getting ready to head up to the main lodge. She told me she had been a nudist for more than three decades.

"There's going to be a lot of shrinkage and headlights down there," she joked, referring to the fact that the temperature year-round down in the cavern was a cool, fifty-two degrees.

I soon discovered at the motel that those attending were a mix of die-hard, card-carrying nudists, and folks like me who were there for kicks, a dare, a once-in-a-lifetime experience.

Pat, fifty-eight, from Virginia, who was partying beforehand at the motel with a group of her cousins, epitomized the latter contingent.

"Why am I here? When I'm in a nursing home laying there in a bed giggling to myself, the nurses will ask, 'Why are you giggling?' she said. "I'll say, 'Oh, I'm just remembering all the s—I used to do.'"

When I arrived at the main lodge, there was a long line to get in. While in line, I met Keith, forty-six, and Michelle, forty-three, from Binghamton. They told me they were divorced, but still enjoyed each other as friends.

Michelle said she had Keith come over to her place to "rehearse."

"He took off his clothes, I took off mine. We said, 'Ok, we can do this,' and put out clothes back on," she said.

Once inside the cavern's main lodge, I was handed a bag to put my clothes in, along with a black, cotton/polyester bathrobe with an emblem on the chest that read, "I disrobed at Howe Caverns Naked in a Cave."

I then went downstairs to the men's room, took off all my clothes, put them in the bag and left it upstairs with a cavern employee at a designated table. At that point, I was only wearing the bathrobe with nothing underneath.

I noticed a long line of folks waiting for a drink at a bar in the nearby room. It was a social hour with two topless female bartenders serving up the drinks. I had a gin and tonic to steady my nerves.

Meanwhile, Sarah Danser, from the Discovery Channel's popular TV reality series, *Naked and Afraid*, was in her street clothes, signing autographs and posing for pictures at a table in the main lobby.

I stopped by and told her I was a big fan of the show. I asked if she had any tips on surviving the fifty-two-degree temperature.

She told me to keep my arms by my side so that no body heat could escape from my armpits, and to cover my chest area with my hands and to run in place. Sounded like good advice.

Now, to the part about Figura getting completely naked. He continues:

New Owners, New Outlook

They started calling us toward the elevators, where we assembled in small groups of about twelve. The elevators took us down 156 feet to the cavern. Once off, you shed your robe and handed it to workers at a table, where they put your robe on a numbered hanger.

The shock of being completely naked wore off quickly. There were bodies of all shapes and sizes. I noticed a wide variety of tattoos.

April Islip, then general manager of Howe Caverns, told me earlier that the group contained individuals ages eighteen to eighty— and that the eighty-year-old was there to celebrate his birthday.

The coolness of the cave didn't bother me at first. I just started walking along the cave's lighted brick pathway, taking in all the interesting geological formations, and cutting an eye occasionally at folks passing by.

On a few occasions, I got a slight chill when drops of water from the ceiling fell on my skin. Following Danser's advice, I kept my arms close to my side and jogged in place a few times. It worked.

Along the path, Howe Caverns employees in bathrobes held up humorous signs reading things like, "Happiness isn't size specific" and "A wee bit nipply." Islip had told me that initially the event was going to be called "Nips and Nubs," but that was dropped because of the sexual references.

I soon came to a table alongside the path covered with small cookies, cupcakes and other pastries. In addition, hot chocolate, coffee, and tea were available.

Nearby, on the other side of the path, was a woman dressed in a gown playing a harp. While I was there, she plucked out the tunes "What a Wonderful World" and "Country Roads."

And farther on down the path was another attraction. Folks could have their naked picture taken for ten dollars. Adding to the fun was a table with items that people could don for their photo shoot, such as a Viking hat with horns, or a bow tie. Those who had pictures taken were handed the memory chip from the photographer's camera that could be used to make a print later, up in the lobby, that they could take home.

After having my picture taken (I didn't use any props), I took the boat ride on the narrow underground lake. I ended up sitting next to Danser and we continued to discuss her *Naked and Afraid* TV show experiences. Small towels were handed out to sit on since the boat's seats were cold.

I spent a total of an hour and fifteen minutes completely nude in the cavern.

Once upstairs, I retrieved my bag of clothes and got dressed. I was handed a medal attached to a yellow strap that read, "You're a sexy beast! Howe Caverns Naked in the Cave 2018." (Today, the medal is draped over the side of the mirror on my bedroom dresser.)

The majority of the post-event conversation in the main lobby consisted of folks saying the totally nude thing wasn't as big a deal as they had previously thought. And, of course, there were numerous joking references to the coolness of the cavern and how it affected one's body.

Michelle, forty-nine, of Port Crane, New York, was philosophical about her experience.

"I felt like we shouldn't have been nervous at all," she said, "because when everyone is naked, nobody can judge each other. The world would be a nicer place if everyone was naked," said Michelle.

Now that we have the reader's attention . . .

Bob Holt, now 65, of Cobleskill, has worked at nearly every job at the caverns since starting there as a teenager in 1971, washing dishes in the caverns' restaurant. He was then a tour guide, worked the ticket office, a clerk and manager of the motel, and resident caretaker for a number of years before moving on. He was brought on as the first general manager

New Owners, New Outlook

Robert Holt of Cobleskill, the first general manager of the cave under Wright-Galasso, Inc. ownership. Holt oversaw many of the first investments the corporation made to attract new visitors.

under Wright-Galasso. Holt is well-known in the hospitality industry, both statewide and nationwide, and in the business side of cave ownership.

At the time of the sale, many of the employees knew Holt, and many of the longer-term employees had worked with him. About a month prior to the sale, he met with key staff. "Several offered excellent ideas for needed change; a couple others were not going to accept a new owner and we knew ahead of time they might not stay with us," said Holt. But overall, he said, both year-round and seasonal staff were excited about the new owners' long-range plans.

"My number one focus was to get the cave attendance numbers back up to a respectable number," Holt wrote in a February 2021 e-mail. To start, a new ad agency was brought on, and a new logo (still in use) was created.

The Howe Caverns restaurant and snack bar was transformed into an upscale café for guests who, Holt said, "were looking for 'grab and go' items of good quality." A popular fudge shop in the lodge became a more expansive candy shop, adding homemade truffles, old-fashioned candy, and other treats.

One of the first things the new owners did to breathe new life into Howe Caverns was to breathe new life into Lester Howe. The animatron describes for visitors how he discovered the cave in May 1842.

Efforts to add life to the old Howe Caverns brand included adding life to Lester Howe as well.

In mid-February 2011, Howe—or rather, a lifelike, animatronic version—began greeting visitors in the small, pre-tour waiting area. At first seated in his library, he tells visitors how he discovered and then explored the cave all those years ago. The animatron was officially welcomed as part of the caverns' attractions in an April media event.

"Untold hours went into creating a show that is historically accurate and entertaining," said Holt, who probably knows more about the Howe history than anybody. "We researched his speech patterns, hair color (red) and every detail to make this an extraordinary experience for our visitors before they enter his treasured cave."

Behind Howe, as part of the display, is a wall-mounted map of the cave and the fanciful names given its formations. (These were described in the final chapter of Section I, "Going Underground.")

"Lester was an instant hit with young and old alike," said Holt.

Unfortunately, the pace and pressures of the twenty-first century were apparently not meant for the nineteenth-century Lester Howe. A piston in

New Owners, New Outlook

his neck has malfunctioned and compressed air spritzes out if turned on. "It looks like he's having seizures," a guide said. Visitors still hear his account of the cave's discovery, though.

The animatron was created by Garner Holt Productions, Inc. (no relation to Bob) of San Bernardino, California. Since 1977, the firm has created hundreds of animatronic people, dinosaurs, animals, sets, and props for such big-name clients as Disney and Universal Studios.

Holt and the Galassos met the California company at the annual International Association of Amusement Parks and Attractions trade show in Orlando. Show vendors provide "anything imaginable" for the amusement industry.

The caverns' contingent began attending the show yearly. That's where plans came together for "Howe High Adventure." The seasonal, youth-oriented attraction includes the ropes course, zip line, and rock-climbing tower, along with a small food concession and a pavilion for picnics and larger events. The amusements trade show is also where they found the H2OGo ball and a company to expand the cave's small gemstone mining feature.

Architect Clemens McGiver (for more on McGiver, see Section III, Chapter 2) was hired to come up with a building that mimicked an old mining facility. The 6,000-square-foot "HC Mining Company" includes both indoor and outdoor sluices for panning gemstones, and a rock and fossil shop. The Sandy Creek Mining Company of Fostoria, Ohio, created the mining experience.

According to Holt, the new attractions and upgrades began to revitalize the cave. Paid admissions went from 120,000 to 150,000 over the next few years under the new owners and management.

Holt left the caverns in late 2012 and became executive director of the National Caves Association (NCA). He oversees day-to-day operations and plans annual events for the 90-plus member caves across the United States, Barbados, and Bermuda. The NCA was founded in 1965, and Howe Caverns was a founding member.

Owning a modern cave brings modern problems.

Maybe a visitor on or about July 7, 2015, took the expression "have a ball" at Howe Caverns a little too literally.

In the following days, Howe Caverns began offering a $1,000 reward for information leading to the return of one of its H2OGo balls, a 12-foot-diameter plastic ball stolen from the hillside of the caverns' estate.

The *Daily Gazette* of Schenectady on July 10 reported that video surveillance showed that the clear-plastic ball was taken late at night by two people on ATVs, though the suspects couldn't immediately be identified from the footage. The case of the missing ball, described as a "self-contained water slide," was turned over to State Police.

"It's got to be local, because they were on ATVs," Caverns Manager William Gallop told *The Gazette*. "So hopefully this thing turns up or we can find out who did it, and we will press charges if that's the case."

Giant clear-plastic balls "stand out," Gallop noted. "If someone were to use this somewhere, people would notice that this ball doesn't belong there." The balls can be deflated for transport, but still weigh about 250 pounds each.

The H2OGo ball ride can place up to three adults inside the ball, along with some water, and then gets rolled 1,100 feet down the front hill. Each ball—there were four originally—costs the caverns about $10,000 each, so the case was considered grand larceny.

At this writing, April 2021, the H2OGo ball has not been recovered. The caverns' management decided to discontinue the attraction.

The Wright-Galasso purchase may have also helped ease the competitive tensions between Howe Caverns and Secret Caverns, according to dedicated local observers and to *Roadside America*, a fun, quirky, online travel guide to the offbeat.

Faced with an unexpected problem resulting from the common use of GPS, the two caves found a way to work together, and hopefully keep the traveling public happy.

The automotive navigation system often leads travelers along the most direct route, not the quickest or most easily accessed from the main highways. As a result, visitors are led along two-lane back roads through the undulating farm and cave country—with few or no billboards to guide them (and occasionally getting stuck behind a manure spreader). GPS sometimes bypasses the picturesque drive of NY State Route 7 along Caverns Road through Bramanville and past a quaint gristmill and the Iroquois Indian Museum.

The solution draws a gasp and quizzical looks from those who have followed the nearly 100-year competition between the two caves—signs for both Howe Caverns and Secret Caverns now share a *single* signpost at the intersection of Caverns and Sagendorf Corners roads!

Perhaps the nexus of the Cave Country, the intersection of Sagendorf Corners and Caverns Road. This pun-filled Secret Caverns sign reads "Udderly Amazing" and "Save Moolah."

Doug Kirby, one of the *Roadside America* site's founders, artists, and writers, is a caver from the Northeast. The site has an affinity for caves, giant muffler men, oddball museums, and those "mystery spots" where gravity seems to run amok.

"For decades we monitored this subterranean slugfest between competing tourist attractions within a mile of each other," the site proclaimed after the sale, tongue firmly in cheek. "There is now peace in the rolling hills of Schoharie County.

"For us, Howe Caverns has always been the preppy fraternity when compared to the 'Animal House' down the road, Secret Caverns," the Web site opined. "Secret Caverns' creative billboards and trippy tours were, of course, no match for the marketing prowess of Howe, but we always thought Secret had the heart.

"We always recommend area visitors check out BOTH caves," reads the online account. And for what it's worth, the author of this book does as well.

At the time this was written (December 2020), many of the caverns' trips and facilities were either closed or offered to small groups because of the COVID-19 pandemic. It is always a good idea to check the caverns' Web sites before going; some tours require reservations.

6

New Owners, New Outlook II

What Almost Was

Joseph H. Ramsey and the Howes Cave Association had big plans when they finally took full control of Howe's Cave in the mid-1870s. Gas lighting was piped from the cave entrance to the head of the underground lake; fallen limestone blocks from the cave were laid into stairways and rudimentary bridges; wooden hand railings were added; and the guides and guests were provided with more fashionable attire for going underground (see the photos by S.R. Stoddard). And, when the association added the elegant three-story Pavilion Hotel to the stone Cave House, they envisioned a popular summer resort that would be "first class in every respect," with all the fashionable accoutrements of the period—bowling, billiards, dress balls, gas lighting throughout, and all the latest "sanitary arrangements."

Lester Howe's cave, they must have reasoned, needed to be more than just a cave tour and an overnight stay.

Nearly 150 years later, new owners Wright-Galasso, Inc. put together plans for new attractions and accommodations that dwarfed Ramsey's efforts in both scope and scale. There's a lot that can be done on the caverns' well-manicured, 330-acre estate. While the accomplishments to date in 2021 have breathed new life into the caverns' business model, two proposals that fell by the wayside would have changed Howe Caverns—and Schoharie County—forever. Locals can argue the pros and cons of these plans—and possibly will—for years.

First was "Dinosaur Canyon," proposed in May 2010 as part of a multi-attraction package that included several projects, such as the Howe High Adventure with climbing rock wall and zip line, that actually were completed the following year (see the previous chapter).

Dinosaur Canyon was envisioned as a 120-foot-deep prehistoric canyon swarming with full-sized animatronic replicas of such species as *Diplodocus, Triceratops*, a flying *Pteranodon*, and *Tyrannosaurus Rex*. A "fossil walk," where visitors could find and pick out fossils from the dinosaur canyon, was planned along with a visitor's center, ticket sales and, of course, souvenirs.

At the time, attendance at the cave was on the upswing, said then–General Manager Robert Holt, by about 13,000 over the previous year, bringing the total up to more than 160,000. "People are staying closer to home," Holt said, "and one thing we always hear is, 'What else is there to do?' People will stay longer if there is more here."

Also part of the package was a 22,000-square-foot entertainment building with an arcade, 4D theater, flight simulators, and a "dark ride" that would move guests through multiple dimly-lit scenes, often animated, to create a theme.

Dinosaur Canyon would have been located east of "Discovery Drive," southeast off the road that leads visitors from Caverns Road to the entrance lodge. (Cattle grazed there for decades; maybe some still do.) The entertainment complex and arcade would have been located to the north of the lodge, once a small picnic area and forest land.

And there was more, reported the Cobleskill *Times Journal* under the headline, "Howe Caverns Unveils Huge Plans." A lot more.

If "future demand and economic conditions" were right, said Holt, a 50,000-foot indoor water park and attached 250-room hotel were possible. The caverns' owners planned to build on the western portion of the estate, constructing a new access road to the caverns from Sagendorf Corners Road. The massive project, Holt told the town planning board, would require a new water system, a new sewer system—including a sewer plant—and new electric and communications systems.

Dinosaur Canyon was expected to create 300 new jobs on site and an additional 450 jobs at local businesses. The hotel and water park would have meant another 350 full-time jobs.

The costs were not made public, but the new owners seemed willing to spend to get visitors in the front door.

Schoharie County has a population of nearly 31,000 and a workforce of 7,800 at peak employment.[1] It's part of the Albany-Schenectady-Troy Standard Metropolitan Statistical Area, and likely would have had to draw workers from there to meet this sudden jump of about 10 percent in the local workforce.

Plans were set aside for Dinosaur Canyon and the indoor water park/hotel resort sometime in early 2014 when the state's gaming commission announced that three new casino licenses would be offered in upstate New York as a way to stimulate the local economies.

The headline of a March 18, 2014, article in *The Daily Gazette* of Schenectady got it right—"Howe Caverns wants piece of the action."

No matter how much kids love them, dinosaurs just didn't show the sort of financial promise the company originally hoped for, a spokesman for the Galassos said. There was also a dispute, apparently, with the company that made the prehistoric animatrons.

The caverns' co-owner Emil Galasso said he laughed, however, when it was suggested to place a casino at Howe Caverns. Not that it wouldn't attract business, but it seemed highly likely that the small rural community would be passed up in favor of a more cosmopolitan setting. "I figured Saratoga Springs had it tied up," he said.

But Saratoga officials didn't want a casino in the city famous for its world-class horse-racing venue, which already offers plenty of betting opportunities.

So, while casino developers—some big, glitzy names in the business—were looking at Capital Region locations for the best chance at landing a state casino license, Galasso began marketing two separate sections of the Howe Caverns estate, totaling 330 acres, as a potential site. Thus was created the Howe Caves Development Corporation (HCDC), Emil Galasso, president. (Grammarians and historians may cringe at yet another derivation of the simple elegance of the singular possessive, Howe's Cave.)

Long-time Galasso employee Chris Tague, now a state assemblyman representing the district, told the *Gazette*'s reporter, "We're not looking to get into the casino business. We want to keep the cavern separate. We'd just sell or lease the land."

A lot of the siting work, engineering, and environmental reviews had already been completed as part of the plans for Dinosaur Canyon and the water park resort/hotel. While the dinosaurs had died off, the water park and hotel could be folded into the larger casino plans.

"We have a lot to recommend a casino right here," Galasso told the *Daily Gazette* reporter. "A developer could just go and get a building permit and start working," he said.

On April 24, the *Times Union* of Albany reported the caverns' development corporation had paid the $1 million application fee to the state's gaming commission, though a casino operator for the Howe Caverns site had not been hooked.

Galasso hinted, though: "Today, we are sorting out—who is in and who is out. When the smoke clears, we are confident that operators will recognize the tremendous potential of our shovel-ready site." They didn't have long to find one, however. A June 30 deadline loomed to submit proposals to the state's gaming commission.

It's not known how many developers called on the Howe Caves Development Corporation with interest in the caverns' property, but Galasso and company settled on Michael Malik, a Michigan casino developer, to finance and own the project; Full House Resorts, which is headquartered in Nevada and lists former Chrysler Chairman Lee Iacocca as one of its founders, was brought in to run the hotel/casino.

"The first time I saw the site, I knew it was a winner," Malik told the Cobleskill *Times-Journal*. "I've developed many resort destinations before, and this site has all the ingredients to be a successful destination resort." That Howe Caverns was already attracting 160,000 guests annually was a plus, they reasoned.

Malik and his business partner, Marian Ilitch, had previously put together three casinos in Michigan. Ilitch and her husband Mike had founded Little Caesars Pizza and own the Detroit Tigers and the Detroit Red Wings. The *Times-Journal* reported that architect Joel Bergman, who had helped design casinos such as Caesars Palace and MGM Grand in Las Vegas, was brought on board as designer. It was a far, far cry from the Ramsey organization's late-nineteenth-century Pavilion Hotel with its 100 rooms, 3 stories high, lit by gas.

The Malik organization referred to the $450 million behemoth as a "transformative opportunity" in the executive summary submitted to the gaming commission. "This project represents an opportunity to effect a game-changing re-positioning . . . [of the upstate region] . . . that for too long, had been an underperforming pocket of New York State."

The Howe Caverns Resort and Casino would "put the entire region on the psychological map of not just New Yorkers, but a target audience well beyond," Malik and company boasted.

"At first, residents were wary of the casino and gambling here," said Jim Poole, owner/publisher of the Cobleskill *Times-Journal* newspaper. "This is a conservative area." He noted that proposals for off-track betting parlors in Schoharie County have consistently been rejected by county officials.

"But as the process played out and people heard more about the plans from the Galassos, they seemed more to cotton to it, to see possibility for real economic improvements, jobs, and tax benefits," said Poole.

The only concern that remained throughout the process, said Poole, was for the world-famous cave. "People were worried about what would happen to the cave."

Page after page of descriptions, legal documents, environmental studies, and architect's renderings were submitted in a binder of nearly 1,000 pages to the gaming commission. Both the Malik organization and Howe Caves Development Corporation were optimistic. It wasn't long before they were said to be among the top sites in contention.

Developers claimed the HCDC would generate as much as $500 million locally in related construction revenue and employ more than 3,000 in construction jobs. When completed, HCDC, they said, would employ 1,700 people, mostly in the service areas.

HCDC planned to purchase 110 acres at Howe Caverns for the project they described, which consisted of:

The casino and 254-room hotel with
- as many as 1,500 slot machines and 60 gaming tables
- three full-service restaurants; convention/banquet facilities
- a pool and spa

The indoor water park/hotel, proposed in 2010, with
- a 250-room hotel
- 55,000-plus-square-foot indoor water park
- nearly 1.5-acre outdoor seasonal water park
- three full-service restaurants
- arcade, game, and entertainment park.

Sixteen applications were submitted for the three casino licenses. Howe Caverns was not among the winners announced that December. The state's gaming commission awarded casino licenses to operations in Sullivan County, in Schenectady in the Capital Region, and in the town of Tyre in the Finger Lakes. Those proposals, the commission later said, drew praise for their financing plans and architecture, as well as their ability to maximize dollars for the state and their local communities.

A headline in *Politico/New York* on February 27, 2015, explained, "Losing casino bids hindered by debt, bats, bankruptcy." Readers might be surprised to learn that it was not Howe Caverns that was taken out of consideration because of bats—it was the proposed Nevele Resort in Ulster County, eight miles from a known bat habitat, whose proposal raised questions about whether that habitat could be preserved.

In its succinct explanation of the gaming commission's decision, the state board saw "a critical concern" that the Malik proposal provided "no commitment or highly confident letters for either its equity or debt financing." Further, the board said, the Howe Caverns proposal stated that it could not propose a capital structure in any level of detail prior to receiving the gaming license."

Additionally, Howe Caves' proposal for a resort at the site of the cave theme park in the Capital Region lacked "credibility," the report said.

Everyone recognized that it was a gamble going into the process, and those going into the casino business should recognize that more than most. There were no cries of outrage from the losing developers, or heavy sighs of relief from those who had opposed the plan from the start.

Although new attractions and new underground adventures have been added to bring in more visitors, Howe Caverns is still, basically, a cave—a remarkable, awe-inspiring, and intriguing cave in upstate New York with a lot of history and a lot of future.

---

1. NYS Department of Labor's Quarterly Census of Employment and Wages report.

The Cave House Museum of Mining and Geology

A Love Affair with Underground Empires

Like this book, the Cave House Museum of Mining and Geology (CHMMG) is envisioned as a showcase for everything that is remarkable about Schoharie County's cave country—its underground empires and the people who made them, their history, geology, and industry.

Saving the Howe family's long-abandoned Cave House Hotel had been the dream of many cavers and historians ever since the Penn-Dixie quarry closed, but efforts to acquire the property or have it placed on the National Register of Historic Places never reached fruition.

It took Clemens McGiver's vision for the "adaptive reuse" of the quarry and everything in it to develop the concept for a museum in the old stone hotel. It took Ben Guenther's love and knowledge of the area's geology, paleontology, and history to help McGiver fully realize the opportunities. And it took Emil Galasso's startup and support for the economic engine McGiver saw was needed to get the museum off the ground and to rejuvenate the long-closed quarry.

Soon after Cobleskill Stone Products announced plans for the quarry by leasing the mineral rights, the 20-acre Cave House property was set aside for the museum. The property's owner, Callanan Industries, did so at Galasso's request, and they remain a strong supporter of the museum project.

The Cave House Museum of Mining and Geology

Things moved quickly at first; interest and support came from seemingly everywhere. It was a rare weekend when someone from the community didn't stop by to learn more and peak around the old hotel. The caving community volunteered to help clean up the property; former quarry employees and older residents of the Howes Cave hamlet donated photos and other mementoes. Several photos held privately for years are included in this book.

A ten-member board was organized to guide the project and solicit funding for the nonprofit organization. The original board consisted of: Michael Galasso and Clemens McGiver, both of Cobleskill; Ben Guenther, of Decatur; Dana Cudmore and Robert A. Holt, of Cobleskill; Judy Johns, of Richmondville; Charles A. Stokes and Maryann Ryder, representing Callanan Industries of Albany; John Finnerin, of Randolph, New Jersey; and Craig Watson, of Cobleskill.

Guenther, as caretaker of the site, and McGiver, along with the Galassos, were tireless promoters of the project. The regional news media tracked the project, as did numerous national trade publications, which were happy to see an organization that would promote the positive side of their industry. "Reincarnation of a Quarry: An aggregate producer hopes to combine quarrying with a museum and education center—all on a historic site," from the May 2004 *Pit and Quarry*, was typical of the industry headlines. Readers are excused for not having seen this or similar stories in *The Low Bidder*, *North American Quarry News*, and *Rock Products*, but it was big news industry-wide.

The first and most urgent step was to save the Cave House from further ruin. For more than a dozen years the building's interior had been subjected to rain, wind, and related weather damage following the 1989 tornado that tore off portions of the original roof of slate. (Dr. William Kelly, curator of geology for the New York State Museum in Albany, determined that the roof—completed during Joseph H. Ramsey's ownership—was of Pennsylvania slate. McGiver added, "When you own the railroad, bringing in a carload of slate from hundreds of miles away doesn't sound like such a big deal.")

A $90,000 grant, administered by the New York Power Authority, paid for the building's energy-efficient doors, windows, and lights. The door and 47 tall Victorian windows were replaced or restored by Contractors' Millwork of Sharon Springs, a unique father and son company that uses historic equipment at their shop, most of it belt-driven.

McGiver described the Cave House interior as "remuddled." Much of the work to get the interior back to its original floor plan and bare frame walls was performed by volunteers from the caving community. Workers uncovered a high interior door, a fireplace and mantel, and other original late-nineteenth-century features that had been covered by new drywall, and painted or wallpapered over. Most of it had suffered badly from years of exposure to the elements.

This is looking out the historic entrance to Howe's Cave, taken soon after some restoration work began at the site. Pictured is Cobleskill Stone Products employee Gary Lull.

The Cave House Museum of Mining and Geology

As they broke through one first-floor wall, workers found the final resting place of hundreds of bats, their bones enough to fill a five-gallon bucket. It seemed likely they hibernated in the dilapidated hotel over the winter months, tucked between the exterior stone and framed walls.

The work to clear the cave entrance began as well. Removing decades of accumulated mud and surface soil collapse revealed a passage of about 30 feet in length, 6 feet high and about 4 feet wide. Just inside, a slightly smaller passage twists around a corner and down to the entrance to Barytes Cave, discovered in the early 1900s—a rarity, a separate cave system beneath another.

At the end of the entrance passage is the Lecture Room. It is here that Lester Howe paused to describe his cave and the rigors ahead on the 8-hour tours. It is a large room, 20 feet wide by 60 feet long and about 20 feet high, and at one time filled with rubble, or breakdown. In the far right corner is a climb leading to the old Bridal Chamber, where Howe's children were married. Though not heavily decorated with speleothems—cave formations—it is awe-inspiring to stand in a portion of the cave seen by so few over the last century.

After the entrance passage had been cleared and lights had been temporarily strung, work began in the cave, too. A compact, walk-behind skid steer was lowered into the pit and maneuvered into the Lecture Room,

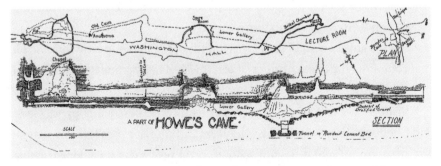

This 1906 map shows a portion of Howe's Cave that was destroyed by quarrying. The Lecture Room, shown at right, is part of the tour at the Cave House Museum. The cave to the left, beyond the lower tunnel into the Rondout cement bed, is either inaccessible or lost forever.

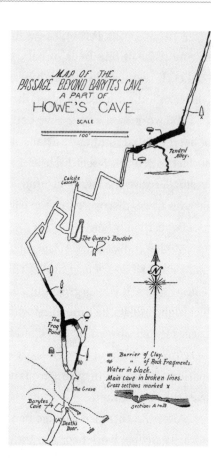

Prof. Cook's 1906 map of Barytes Cave, the entrance to which lies beneath the original entrance to Howe. Note the passage names near the entrance—"Death's Door" and "The Grave." Much more has been added to the cave, which is part of an extensive drainage system lying to the north that includes the Secret-Benson Cave system, Wolfert's Cave, and others.

where Cobleskill Stone Products employee Gary Lull began moving and clearing spots to walk and explore.

A wrought-iron set of stairs was found on site that fit neatly into the pit that contained the cave entrance; it was lowered in by crane lift and seemed to fit perfectly.

Visitors started to come, slowly at first, invited personally by McGiver and Guenther. They were led by volunteers and consisted of a few classes here and there from Capital Region schools, or of professional geologists being given a wide-open outdoor lab to show students. Kevin Berner, a professor in the Fisheries and Wildlife Department at the State University of New York at Cobleskill, brought students to tour the cave and mine with Guenther to count and identify bats and explore the area's ponds and habitats.

The Cave House Museum of Mining and Geology

"The quarry is a wonderful outdoor laboratory with outstanding exposures of the Silurian and lower Devonian rocks of the Helderbergs that are more than 400 million years old," said Dr. William Kelly of the New York State Geological Survey. "In addition to the excellent stratigraphy, the mine contains well-exposed examples of fault and karst [cave] development ... It provides a great location to study geology in the field."

Guenther had made numerous contacts with the New York State Museum and the State Geological Survey in Albany well before the CHMMG was begun. Kelly, the State Museum curator, was a frequent visitor and helped spread the word about the Cave House to other enthusiasts around the state.

Kelly and others at the State Museum were strong allies that helped the museum board secure its status as an educational institution. It was a huge accomplishment for a young museum that was not quite open to the public at that time.

Ben Guenther in the water crawl in Barytes Cave, which lies beneath the original entrance to Howe's Cave on the grounds of the Cave House Museum of Mining and Geology. Photograph courtesy of Paul Rubin.

In November 2003, the Cave House Museum of Mining and Geology was awarded its provisional charter by the Board of Regents of the New York State Board of Education, thereby elevating its status as an educational organization and formalizing its long- and short-term educational objectives. The charter also allows the museum to pursue closer ties with the State Museum, which includes the sharing of mineral displays, interpretive tours, and more.

Other exciting things were happening at the site, as well.

A granite monument created by North American Cement Company to honor its miners had been moved by the county's Industrial Development Agency from the quarry to another large industry in Cobleskill. It was brought back and now graces the lawn adjacent to the quarry entrance. Galasso called it, "Proof that the plan for the quarry was meant to work. If it had stayed at the quarry, it would have been vandalized. It was preserved [at the Cobleskill site], now it's coming back," he said in the December 2002 *Low Bidder*.

Fund-raising efforts, that include a yearly golf tournament for quarry industry professionals, got underway. The golf tournament, which raises

Mr. and Mrs. Charles E. Dewey Jr., of Battle Creek, Michigan. Charles was the great-grandson of Lester Howe. The family has always felt a connection to the cave; Dewey donated $500 to the Cave House Museum when he heard the old family hotel was being turned into a museum. He passed away in 2014.

The Cave House Museum of Mining and Geology

as much as $40,000 each year, is still the museum's major funding source. Quarry professionals from around the region pay $3,000 per foursome. There's also now an annual "Gem, Mineral, Fossil and Jewelry Show" held the third week of September. The location has varied, but organizer Nancy Sagendorf said the show brings in "a few thousand" each year.

A descendent of the old hotel's original occupants contributed as well. Soon after plans for the museum were announced, a $500 check came from Charles E. Dewey Jr., of Battle Creek, Michigan, the great-grandson of Lester Howe. Dewey told a friend he considered the donation a birthday present to himself. He had turned 80 on June 30, 1997.

Dewey's grandfather was Hiram S. Dewey, the railroad surveyor who married Howe's daughter Harriet Elgiva in the old cave's Bridal Chamber on September 27, 1854. Dewey was working for the Albany-Susquehanna Railroad at the time as the line wound its way past Howe's Cave. The couple eventually settled in Jefferson City, Missouri.

Dewey (1918–2014) had been following from afar the plans to restore his family's old hotel through his friendship with Robert Holt of Cobleskill. Holt, a former general manager at the cave, was one of the museum board's original directors.

"All these years later, the Howe descendants are still proud of their family's heritage and how it is still enjoyed by tens of thousands of people each year," said Holt. "It was a generous and touching contribution."

Things were also moving forward with plans to open a public park south of the quarry, just across the Cobleskill Creek. Forty acres had been donated by Callanan, also at Galasso's request, as part of the deal to lease the quarry's mineral rights.

The site was once the location of several Tite Nippen homes, two suspension bridges across the creek, and two softball fields built for employees by the Penn-Dixie company. The new park would be named after its opening as the Doc Reilly Park, for Dr. Francis S. Reilly (1933–2004) of Cobleskill.

One way in which park promoters raised money for the new Tite Nippen-area park took advantage of the unique connection between the cave, quarry, and community.

On September 11, 2004, sixteen contestants began the world's first—probably only—"underground duck race." The rubber ducks, sponsored and decorated by local businesses, entered the underground stream at the end of Howe Caverns' Lake of Venus. Organizers anticipated they would gradually float their way 1,800 feet through the undeveloped portion of the cave and emerge in the surface ponds on quarry property.

Sponsoring local businesses paid $100 for their rubber duck and then collected $2 bets on their entry to surface first as the winner. The sponsors got creative. They named their rubber ducks, imprinted them with company logos, and painted them—some intricately, including a "Spider Duck."

No one knew how long the great underground duck race might take. The stream is barely a trickle in that part of the cave; contestants faced sticky mud, clay banks, and crevices as well. Organizers specified that in the event no duck made it from the cave within six weeks, by October 30 the contestant who generated the most in donations for the park would be declared the winner.

This contestant in the world's first underground duck race was sponsored by a local hair salon. The race along the underground stream had unexpected consequences.

The Cave House Museum of Mining and Geology

Guenther became an "official race proctor" and entered the cave to check on contestants a few days after the race began. He reported finding seven ducks that had reached a historic section of the cave named the "King's Corridor." The large, open passage was about the midway point of the race, and from there the cave stream drops beneath the quarry floor and leads directly to the ponds.

Though muddy and haggard, the ducks appeared to be doing fine, Guenther reported at the time, tongue firmly in cheek. Race organizers began checking the ponds periodically for a winner.

Few could have anticipated what happened next.

On September 17 and 18, the remnants of Hurricane Ivan slammed New York's southern tier, resulting in heavy rains, flooding, and a federal disaster declaration. Howe Caverns was heavily flooded, and the commercial portion of the cave was closed to tourists.

Fifteen of the sixteen rubber ducks were never seen again. They were evidently washed from the cave into the overflowing quarry ponds, then on to the Cobleskill Creek, and finally floated north-northeast to the Schoharie Creek. From there the gaily painted children's toys flowed north to the Mohawk River, east to the Hudson River, and finally into the Atlantic Ocean at New York City harbor.

The sole survivor, the contestant entered by the Bank of Richmondville, was found many days later, stuck in a high crevice in the old part of Howe's Cave.

The world's first underground duck race brought in more than $3,100 dollars for the park along Route 7 in Howes Cave. The unique race was conceived by Kurt Willwerth, a lifelong Barnerville resident and member of the volunteer committee hoping to create the park.

While school groups, geologists, cavers, community members, and others had been visiting the Cave House almost from the very beginning the quarry was opened again, it wasn't until the summer of 2007 that there was anything resembling a "grand opening." There had been too much work to be done to repair the old hotel to welcome guests. A lot remains to be done to the building, but its location and history remain a huge draw.

The Museum of Mining and Geology is still very much a work in progress. Most people would not recognize it as a destination for visitors at all. The property is entered off a back road; there are no large signs or billboards; and other than there being an unpaved parking lot, landscaping is minimal. Indoor plumbing is still a work in progress, but porta-potties are available and bottled water is provided.

Howe Caverns was sold in April 2007 to Wright-Galasso, Inc. (see previous chapter). The new owners, at first, saw a lot of potential in the two properties working together. There was even talk of reconstructing the 300+ feet of old Howe's Cave in the quarry to return the cave to its former state.

Cave House board member Holt was named general manager of this new incarnation of Howe Caverns, and he planned a coordinated effort to open the no-frills Cave House museum to the public on the Fourth of July

This is the old entrance to Howe's Cave. The nonprofit Cave House Museum of Mining and Geology offers a limited number of tours into this short section of cave during the summer months.

The Cave House Museum of Mining and Geology

Geologist Paul Griggs, a member of the Cave House Museum board of directors, has created much of the nonprofit museum's educational offerings and leads many of the tours as a volunteer.

weekend. Holt put Rich Nethaway, one of the caverns' longest-serving employees, on the job.

"We got some showcases from a closed museum, along with some the museum already had in boxes," said Nethaway. Nearly all the collections had been donated to the museum from former residents, former quarry employees, the caving community, and others who supported the project. Some large rock specimens came from the New York State Museum, hauled to the site by Cobleskill Stone Products equipment. It is a diverse collection, small, yet representative of the museum's mission.

Currently the museum is operated entirely by volunteers. Admission is free, but donations are gratefully accepted. Self-guided tours of the museum are generally available between Memorial Day and Columbus Day on Saturdays and Sundays from 10 AM to 5 PM, but hours may vary. There's no phone service yet, but they do have a Facebook page. Check before making the trip.

Guided tours of the Cave House museum, quarry property, and Cobleskill Stone Products' operation are available by email: pgriggs @nycap.rr.com.

Donations may also be sent to the Cave House Museum of Mining and Geology, P.O. Box 220, 112 Rock Road, Cobleskill, New York 12043.

It's clear that creating the museum is a labor of love. Plans are necessarily big to accommodate the museum's educational goal, but the budget remains small.

The small museum building has an important story to share. Its displays and volunteers describe the nearly 200-year history of Howe's Cave; the underground mine on site that quarried limestone for natural cement in the 1860s; a century-old surface quarry used for manufacturing Portland cement; the site's modern, high-tech, environmentally sound quarry for crushed stone products; and the geology of the area and of New York State. In a sense, they're trying to describe and interpret all that's included in this book.

Prepared by the museum's volunteers, a portion of the historic tour is described here. Visitors find themselves in many of the locations described in Section I of this book and can imagine what it was like to visit Howe's Cave from the mid-1800s until its flashy reopening as Howe Caverns in 1929.

From the volunteers' presentation:

> Guests were welcomed at the Cave House by, perhaps, Lester Howe himself, or by one of his family. They could feel cool air from the cave circulated up through the hotel on hot summer days, and crystal-pure water from the cave slaked visitors' thirst and kept food fresh.
>
> Visitors came from all over the world and stayed in this four-star hotel. People would arrive by horseback, in wagons or, after about 1870, by train. Eventually, the hamlet of Howes Cave grew around the base of the hill and had its own train station.

From the porch of the Cave House, today's visitors can still get a glimpse of the hamlet's once-prosperous main street.

Cave visitors used to walk up the hill from the train station and enter the cave through a grand tunnel of laid-up stone that is currently being

The Cave House Museum of Mining and Geology

renovated; or, if they stayed overnight at the hotel, they would enter the cave through the basement. The "grand tunnel," had been hidden for decades, covered, and filled to accommodate roads built for heavy quarry equipment. It was uncovered in the last decade as work to reopen the cave's original entrance got underway.

Volunteers describe the different phases of the Cave House, which culminated in the huge addition, the Pavilion Hotel, which was destroyed by fire in 1900. The Pavilion extended over what is now the visitors' parking lot. The stone Cave House remained, later becoming a boardinghouse where contractors working to develop Howe Caverns stayed in the mid-1920s; then it became offices and a chemistry lab for the cement companies that worked the quarry over the years.

Some of the displays and the presentation describe the original underground mine. The natural cement mine, started in 1867, extends beneath a large portion of the old Howe's Cave passage, the surface mine, and hamlet.

Another exhibit highlights life during the heyday of the Howes Cave hamlet—a typical mining town—from about the 1930s to 1950s. There's an ore cart similar to that which would have been used underground, a clay model of the area, and displays highlighting the local mining industry. (The remarkably detailed clay model is from McGiver's work on his architectural thesis for adaptive reuse of the quarry.) Another exhibit offers information about the biology and ecosystem of the cave and surrounding rural area.

A short video about the history of the property and plans for the Cave House is offered.

The museum exhibits feature geologic samples collected locally and from around New York State. Just outside is the Yngvar Isachsen (1920–2001) Rock Garden, named for a 43-year member of the New York State Geological Society, educator, and author of *Geology of New York, a Simplified Edition*.

Paul Griggs, of Brunswick, east of Troy, joined the Cave House Museum as a trustee in 2010 after working with Ben Guenther, whom he'd been introduced to by quarry owner Emil Galasso in 2004. Griggs, a licensed professional geologist, and president and principal geologist of Griggs-Lang Consulting Geologists and Engineers, began volunteering weekends and helped developed much of the educational material used today.

"Ben and I shared many interests," said Griggs. "I was interested in the educational outreach chances and I saw the Cave House as a good way to expose students—of all ages—to geology, related sciences and the mining industry."

Griggs's tours now attract 1,500 to 2,000 guests a year. "Most of our tours are for high school earth science classes, but we've had gem and mineral clubs, boy scouts, girl scouts, cub scouts, senior citizens' groups, college classes, mining groups and geology groups, to name a few."

Tours are divided into small groups to explore the Cave House interior exhibits and/or descend the wrought-iron staircase to the Lecture Room; they see and learn about the rock outcrops around the site and then walk through the rock garden. "Most of the schools have prepared a scavenger hunt," said Griggs. There are fossils to identify, evidence left behind during the Ice Age, and different features of cave development and formations to discover.

Griggs's tour can be as simple or complex as needed. His geology hikes take visitors back in time millions of years. There's a lot to be learned here.

The geology-centered tours describe:

- the history of the cave and how caves are formed
- how the Ice Age and glaciers shaped the landscape, and the clues they left behind
- mining history, from the mid-1800s to today's high-tech stone crushing operation on site
- how limestone products are used in everyday products such as paint and aspirin, as well as for roads and heavy construction projects
- and the roles science and geology play in modern mining.

The Cave House Museum of Mining and Geology

"Geologists read rocks the way people read books. They are scientists that study the earth and the processes that shape it. By observing these current processes and applying these observations, geologists can tell what happened thousands and millions of years ago and predict what is lying below our feet," said Griggs.

Sadly, Ben Guenther didn't live to see the Cave House Museum of Mining and Geology become all that its many advocates envisioned. Benson Pruefer Guenther, 70, of Decatur in Otsego County, died March 12, 2008, at home surrounded by his family after a courageous battle with pancreatic cancer. He was born August 29, 1937, in Providence, Rhode Island.

Ben was an avid caver and a member of the National Speleological Society. In the last several decades of his life, he studied geology and paleontology and was eventually drawn to the Howes Cave quarry. There, he met and became lifelong friends with Clemens McGiver.

"Ben was truly a Renaissance man," said McGiver, "a prince among men." He was an accomplished pianist, a painter, sculptor, and photographer. He wrote poems and short stories.

Ben knew the quarry, its caves, and its history better than anyone.

From his obituary in *The Oneonta Daily Star*:

Ben initiated a geological exploration of the immediate area and encouraged educators to see the quarry, underground and surface mine as a natural history laboratory. His ideas took hold and with great enthusiasm he received endorsements from the New York State Museum, colleges and universities, high schools, elementary schools, and numerous other educational groups. They visit the quarry today because of Ben's efforts.

Through Ben's determination, self-gained knowledge, and practical application of work at the quarry he was invited to join Cobleskill Stone Products of Cobleskill, to take part in one of the most ambitious restoration projects attempted in recent times in the area. The efforts attracted national attention.

Through Ben's tenacious efforts and talent, The Cave House

Benson P. Guenther, 1937–2008.

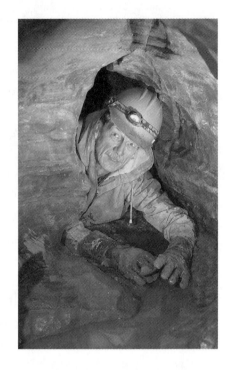

Museum of Mining and Geology became a reality. Ben was invited to serve as a charter member of the Cave House board of directors and served as education director of the Cave House Museum until his illness. He also worked diligently on an expansion plan to reopen portions of the Howe's cave system not seen in over a century.

Ben was survived by his wife of 50 years, Caroline (Bettencourt) Guenther, whom he married Sept. 13, 1957; his son, Benson Guenther, and wife, Sandra, of Oneonta; his daughter Wendy Guenther, of Decatur; his daughter Christina (Guenther) Tavanian and husband, John, of Providence, Rhode Island; and three grandchildren, Jessica Guenther of Oneonta, Aisha Guenther and Quentin Guenther of Decatur; his sister, Ardys (Guenther) Filippone, and husband, Joseph, of Rhode Island; as well as a niece and nephew.

In honor of Ben's passion and devotion regarding education, the Cave House Museum board of directors passed a resolution to create a library at the museum in his name, the Benson P. Guenther Library.

Also at the museum site is the "Guenther Resurgence of Howe Caverns." Ben discovered the original spring resurgence of what is today the River Styx in Howe Caverns, on the hillside above the museum. According to karst hydrologist Paul Rubin, the underground stream flowed one to two million years ago through the slowly forming cave and popped to the surface at the spring, eventually flowing to the Cobleskill Creek to the south. (At the time, the creek was 150 feet higher than today.) The remnants of the spring, in the form of a small cave, offer geologists the best estimate available to date the age of Howe Caverns.

Ben Guenther would be proud to see his dream of a museum at the Howes Cave quarry gradually coming to life. He'd be one of the first to lead tours of the cave, the old mine, or to smile broadly and welcome visitors through the front door. Who knows what else he might have discovered there?

A lot has changed at Howe's Cave, Howes Cave, the quarry, and at Howe Caverns since that curious farmer first pulled back the brush that concealed a hidden empire underground. Other adventurous entrepreneurs followed suit, exploiting the natural resources of the cave country both below and above ground.

There have been exciting new finds since the 1990 publication of the author's first book, *The Remarkable Howe Caverns Story*. New caves have been found, and explorers have pushed the limits to discover new passages in known caves. None may turn out to be as capable of commercial "empire-building," as were Howe or Secret caverns, but hope, like water from the River Styx, springs eternal.

New uses may be discovered for limestone and its chemical composition. Cement, cut stone, and asphalt production continue to be modernized, and bustling quarries operate with new appreciation for safety and the environment.

And Lester Howe's Garden of Eden may still be out there somewhere in Schoharie County's cave country.

Epilogue
Connecting the Cave Community

If the nexus of the Cave Country in Schoharie County is not the Sagendorf Corners Road and Caverns Road intersection, it is certainly a busy crossroads. The Secret Caverns entrance lodge is about a half mile to the northeast on the appropriately named road; there, head southeast on Robinson Road to find the significant, connecting Benson's Cave system and numerous sinkholes. More sinkholes dot the landscape just to the southeast. One of them enters Wolfert's Cave—two, forty-five-foot drops leading to a strenuous, challenging connection to Barytes Cave and the old Howe's Cave entrance.

Walk a half mile southwest across the field from the intersection and the 1802 Sagendorf homestead to the Howe Caverns lodge and the cave's manmade entrance. A half mile down the road, behind the old Van Natten farm, is the fabled "Sinks by the Sugarbush," where artifacts thought to be from Lester Howe's collection were found.

Head southeast down the Sagendorf Corners Road hill about three miles; the growing Howes Cave stone quarry is on both sides of the two-lane road, the quarry walls dropping 100 feet to the floor, or "pit." There, heavy construction equipment rumbles about and the rock-busting crusher jaw pulverizes stone to ever-smaller sizes. Just inside the quarry is the old site of the regal Howes Cave Association's Pavilion Hotel. The beautiful stone Cave House remains and is now a museum-in-the-making. Adjacent to that is the original Howe's Cave entrance and the underground connections to the Robinson Road caves. Cool air still emanates upward on warm days.

Look out across the valley from the Cave House porch. Visitors see Terrace Mountain, with caverns, sinkholes, and springs on all sides. Imagine a straight line heading southeast from the Howe's Cave entrance less than four miles to the other side of the escarpment and to VanVliet's Cave, along Schoharie Creek. That's the "finger of geology." Look just to the right, on the hillside across Cobleskill Creek, and you can imagine

Epilogue

Lester Howe's farm property and acreage that he called the "Garden of Eden." Maybe there's a fantastic cavern hidden there still.

Within walking distance of the Cave House, just south, is what were once the tracks of the Albany and Susquehanna Railroad, now the Delaware and Hudson. There was a depot in the little hamlet at one time, as well as a post office, two churches, a schoolhouse, grocer, luncheonette, and Chevy dealer. You could get a cold beer at Sammy Sautin's raucous Howes Cave Hotel and go square dancing Saturday nights. During the week you could get your haircut there.

Lester Howe's discovery, and more importantly his decision to share it with the world, made it all possible. This underground empire helped inspire profound insights into the natural world for some; it created fortunes and offered adventure and opportunity for others; and above all, it built the richness of life shared in a small community.

This is uniquely cave country.

Little, if anything, is known of the interactions between the historic figures in this book. Howe was a relative newcomer to the area when he first explored well beyond the entrance of the strange "blowing rock." Most of his neighbors had heard the legends of a cave before Howe came upon it and decided to investigate.

The Sagendorf Homestead—pockmarked with caves and sinkholes—had been there at least twenty-five years prior to Howe's arrival. Did they welcome and accept the eccentric new neighbor, or did they think his interest in the underground odd? After the cave became famous, did the Sagendorfs explore the new cave? ("Everybody did!" a relative said enthusiastically.)

It was a second generation of Sagendorfs on the homestead farm when Howe's Cave was at its most renowned. George Sagendorf (1783–1858), one of nine children, and his wife Catherine (1799–1868) raised six children of their own on the busy farm. The Howe and Sagendorf families probably attended the Reformed Church services in nearby Barnerville and other community events.

It is reasonable to suggest that the Sagendorf farm provided meat and dairy for the Howe family's guests, and that the Sagendorf family enjoyed the music at the cave that Howe and his daughters sometimes offered.

Howe had a penchant for fireworks. It is likely that Fourth of July celebrations were popular community events at the cave. It is also likely that the community grieved together, shocked and saddened by the April 1865 Lincoln assassination and other tumultuous national events.

The interest in Howe's Cave died down in the early years of the twentieth century for reasons described earlier, although the need for the limestone resources of the area grew steadily. The quarry's growth eventually resulted in the cave becoming off limits.

But there were other caves nearby just waiting for adventurers. Teenaged "Dellie" Robinson charged a few pennies to lower friends and guests into the caves on his farm property just up the hill from the old cave. The Robinsons were later arrivals to the cave country, buying a farm sometime around 1880. Dellie was only a year older than John Sagendorf, and they lived only about a mile apart. A connection seems likely there. It's easy to picture the two young men, their brothers, and friends exploring the caves throughout that area after farm chores were done. Although about five years younger, Arthur Van Voris might have joined them as well, grabbing a Bright Star flashlight and other gear from the family hardware store on Main and Grand streets in Cobleskill, and heading off to go cavin'.

In 1906, when Prof. John H. Cook visited the caves of Schoharie County for the State Geological Survey, he likely would have spoken with the Sagendorfs and Robinsons. There is an unnamed assistant in Cook's text and in a photo taken deep within Howe's Cave. It's only known that it wasn't John or Dellie, who had typhoid fever around that time, according to a niece.

As engineering advances made it possible to reopen old Howe's Cave, the community buzzed with enthusiasm. Yes, the Sagendorfs hoped to profit from the cave that lay beneath their property, and history shows Robinson was an ardent supporter. One reason the Howe Caverns corporation sold $100 shares of stock was to bring their neighbors into the

Epilogue

opportunity they saw ahead of them. They offered an installment plan for those who couldn't afford the full payment all at once.

Roger Mallery, who led the cave revitalization crews, didn't have the caving bug when he first came to Howes Cave from Binghamton, then in his mid-twenties. He caught it fast, however, exploring about the countryside for a cave to call his own. He found it in Nameless Caverns, just up the road from the Sagendorf farm. He married a Howes Cave girl and settled down, keeping the cave business in the family. He sought other underground empires to create as well.

Van Voris's newspaper articles in the spring of 1929 shared the adventure of the "lesser caves" in the area and introduced them to a wider audience. Along the way he introduced "the VanNatten boys" of the cave country, as well as a young, smiling upstart, Edward A. Rew, nearly fifteen years his junior. Searching for Howe's Garden of Eden, Rew introduced the world to that "Finger of Geology."

The reopening of Howe Caverns in 1929 might seem like the climax of a good adventure tale, but since then there has been nearly a century of cave country discoveries, exploits, adventures both comic and tragic, and ongoing advances in the sciences of cave geology and karst hydrology. A few were described here. There are many rare individuals, some mentioned in this book, who in their own ways have contributed significantly to the rich history of this remarkable little slice of upstate New York. Those histories can be written at a later date.

"The caves, sinkholes, and springs in the area were always a part of their lives," said Nancy Sagendorf, a retired science teacher. "My father [Willard Sagendorf] did some exploring and used to take me with him, starting around the time I was ten."

Nancy's early experiences probing the depths of the cave country offered a bit of wisdom that can be applied to life in general: "He taught me to only go a couple of feet each time, as you never knew what you would find.

"Make sure you had enough energy to get out . . . and then go [a little farther] each time."

Readers interested in sport caving are encouraged to find a caving club in their area and visit the Web page of the National Speleological Society.

Readers interested in visiting a show cave developed for the public are encouraged to visit the Web site of the National Caves Association.

"Take Nothing but Pictures, Leave Nothing but Footprints, Kill Nothing but Time."

—National Speleological Society motto

SECTION IV

Related Tales from the Cave Country

The Gebhards of Schoharie

*Father and Son Pave Way for Earliest Studies
of Geology, Natural History*

During the early decades of the 1800s, in the formative years of the Scientific Age, John Gebhard and his illegitimate son, John Jr., together explored the hills, caves, and river valleys in Schoharie County, a region made remote from the prosperous Albany settlement on the Hudson River by the Helderberg Mountains.

Recording, collecting, cataloguing, and intuitively interpreting their discoveries over a fifty-year period, these methodical amateurs were the first to use many of the scientific principles that would later be incorporated into our modern understanding of the fields of geology, stratigraphy, and paleontology. They were decades ahead of their time. Their cave discoveries, rock and fossil collections, and willingness to share their intellectual gains made Schoharie County a Mecca for the aspiring geologists of their day.

By later moral "standards," the birth of John Jr. on October 22, 1802, was scandalous, but it was not necessarily uncommon in early post–Colonial America. His mother, Elizabeth Bouck, was the twenty-year-old niece of the future New York governor, William C. Bouck (1786–1859), of the neighboring Town of Fulton. His father was from a prominent family that had settled along the Hudson River north of New York City during the Revolutionary War. John Sr. (1782–1854) and his older brother Jacob were among the first to practice law in Schoharie, not long after it became the county seat in 1795. Both brothers rose to political prominence and were active in community affairs at the county and state levels.

For much of Junior's life, the Gebhard and the Bouck families—both John Sr. and Elizabeth eventually married others—lived within walking distance of one another in stately homes on Main Street in the village of Schoharie, an awkward arrangement at best. To further complicate the historical record, a third John Gebhard was born to John Sr. and his new wife in late 1812.

Schoharie County, then and now, offers much to anyone with an interest in the natural sciences. Within its 622-square-mile boundaries are rolling hills, high escarpments, picturesque waterfalls and gorges, fertile river valleys, caves, and sinkholes. There are clear-water and sulfur springs, and other topographic features, fossils, and landscapes that have left behind clues to a geologic past extending back millions of years.

Many of the Gebhards' earliest discoveries were literally in their backyard. An extended ledge of exposed and weathered limestone partitions the entire eastern length of the village of Schoharie and contains Becker's, a small cave. The Gebhards discovered fossils and minerals there; some were

John Gebhard, Jr. (1802–1889), son of a prominent Schoharie attorney and an unwed woman from a prominent political family, became—working with his father—a famous naturalist, cave explorer, and fossil collector in the early to mid-1800s. In 1849 he became the second curator of the New York State Museum; over the years, much of his private collection from Schoharie was added to the museum's vast archives.

among the first known deposits in the nation. East of the village, along the banks of the confluence of the Schoharie and Fox creeks, young people can still find curious rounded "clay buttons" as well as "turtle rocks"—a limestone concretion resembling a turtle's shell.

John Sr. developed a prosperous legal practice. Around 1830 he retired from the bar to pursue his interests in geology. The reasons are not entirely known to historians; however, by that time he had accumulated a sizeable nest egg of $50,000. By then, Junior was also well-established, having been elected county clerk in 1828. He later served as a justice of the peace as well.

Joining the "Father of American Geology"

At first father and son labored in the seclusion of the Schoharie Valley. There were few, if any, experts in the fields of fossil collecting, mineralogy, or stratigraphy, which studies the earth's strata, or layers. By a twist of fate, in about 1830 John Sr. reconnected with longtime childhood friend Amos Eaton (1776–1842), then teaching in Troy on the other side of the Helderbergs and across the Hudson. Professor Eaton had also given up a legal career to devote himself full time to the study of geology, teaching, and writing textbooks. The founder of the Rensselaer School, now prestigious Rensselaer Polytechnic Institute, is widely regarded as the "Father of American Geology," and the period 1818–1836 is regarded as the Eatonian Era.

Eaton brought the Gebhards and their discoveries into contact with other educated men of the period. Two of the Gebhards' findings were first published in the January 1835 edition of the prestigious *American Journal of Science and Arts*, a quarterly publication edited by one of Eaton's former professors at Yale. Nearly a dozen pages are dedicated to two of the father and son's important discoveries—the first full exploration of Ball's Cave on Barton Hill by John Sr. and others, and Junior's discovery of the rare mineral strontianite, used in refining sugar and making fireworks burst red.

The Gebhards of Schoharie

The September 1831 discovery and exploration of Ball's Cave was greeted with great interest worldwide following the *American Journal* article; it was among the first caves in the post–Colonial Era to be explored and documented by educated men.

Historian Jeptha R. Simms's 1845 *History of Schoharie County and Border Wars of New York* gives a thorough and often enthusiastic description of the exploits of Gebhard (and others) there:

> Gebhard's Cavern (called formerly Ball's Cave) ranks conspicuously among the natural curiosities of the county. I have chosen to call it after John Gebhard, Jr. Esq., its present proprietor; a gentleman who has done much to advance the science of geology—particularly that branch now denominated paleontology. This cavern is situated upon an elevation called Barton hill, its entrance being in a piece of woods nearly four miles east of the courthouse.
>
> On the 21st of October of the same year, Doctor Joel Foster, Mr. John S. Bonny, John Gebhard, Esq., and several other citizens of Schoharie, having prepared a boat, again visited this cavern, and being let down by ropes with their skiff, they pretty thoroughly explored it.

After describing the adventurous first and second visits, Simms remarks:

> Gebhard's cavern has a merited celebrity on account of its secluded locality, its limpid lakes, its rotunda, its salubrious atmosphere, and the immense quantity of beautiful minerals it has afforded the admirer of Nature's handiwork; not a few of which, for their snowy whiteness, are scarcely equaled by those of any other cavern in this country: and it will continue to have numerous visitors, although other caves, dark and deep, may become justly celebrated . . .
>
> For as a previous writer observes: "The novelty of navigating a crystal lake by torch light, beneath an arch of massive rock, at the distance of some hundred feet from the surface of the earth—the

breathless excitement resulting from the real and imaginary dangers of the enterprise, &c., are themselves sufficient to render this cavern a place of frequent and interesting resort."

In 1836, John Sr. and village tailor John Bonny discovered, explored, and reported on another cave, found just north of the Schoharie Creek on the farm of Peter Nethaway. (This is now VanVliet's Cave, described earlier.)

Ball's Cave and Nethaway's Cave made Schoharie County famous for its caves a decade before that other local cave was found on the other side of Terrace Mountain, near the Cobleskill Creek.

It was not until May 1842 that that other, spectacular, Schoharie County cave assumed the spotlight—that was Howe's Cave, of course, which still welcomes visitors as a popular tourist destination. There is no historical record of the Gebhards being among the early visitors to Lester Howe's cave, but it seems likely they would have been. John Jr. would have been in his early forties at that time.

Now in contact with other scientists and naturalists, the Gebhards welcomed all to their Schoharie home to share their collections and discuss the latest scientific theories. Eaton brought his students from Rensselaer; others came from Union College in Schenectady as well as from the numerous seminaries in the region.

Eaton's students included men who later made their own remarkable contributions to the field of geology. His students Ebenezer Emmons (1799–1863) and James Hall (1811–1898), among others, became giants in their field.

The preeminent geologist of the day, Britain's Sir Charles Lyell (1797–1875), picnicked at the Gebhard's Schoharie estate in mid-September 1841 during the first of four U.S. tours. Lyell's 1830 textbook, *Principles of Geology*, was the first to introduce the concept that natural, observable causes were responsible for shaping the Earth over unfathomable "Deep time." These causes, erosion for example, continue today, he noted. Lyell's work contributed significantly to Charles Darwin's thinking, and the geologist is referenced in *The Origin of the Species*.

Lyell was most fascinated with the Helderberg Escarpment, and many leading geologists today consider the Helderbergs—which separate the Albany area from the Schoharie Valley—as "hallowed ground." The mountains have served as a field laboratory for geologists for nearly two centuries, studied by men such as Lyell, Eaton, Hall, Emmons, and many others. Now a portion of the escarpment is open as the John Boyd Thatcher State Park, and visitors can walk the same paths the nation's leading geologists did 200 years ago and enjoy beautiful views of the Capital District. A few fossils can still be found in the streambeds there.

The Gebhards shared their collection of "Helderberg fossils" with Lyell, his wife Mary, and Professor Hall, who accompanied them.

In his 1998 book, *Lyell in America*, author Leonard G. Wilson wrote that "Gebhard showed Lyell a slab of what he called 'Tentaculite limestone'—covered with awl-shaped shells. Lyell recognized it as the spine of a sea urchin and advised [the] Gebhards to look for the body of the animal from which they came.

"Sometime later, Gebhard found the giant chambered crinoid bulb . . . which he called 'Lyell's Sea Urchin.'"

The prestige of the Lyells' visit to the rural hills of Schoharie County cannot be exaggerated. It also stirred a great deal of curiosity among the locals—mostly farm folk—wondering why someone would travel days and 3,500 miles across an ocean to look at "a bunch of rocks and animal bones."

Wilson quotes Mary Lyell: "While we were in Mr. Gebhard's collection, a number of different people came in and no one took the slightest notice of them. They just stood staring at Charles without speaking and then walked out.

"It was so odd I could hardly help laughing," she wrote in her journal.

It was a beautiful fall day, and the Gebhards took the Lyells and Hall to visit geological sites and have a picnic. As they were sitting down to eat, there was another memorable event, Wilson reported; "A large hawk swooped down from a tree to seize a chipmunk, so close to them that Hall tried to catch both. The chipmunk escaped and the hawk lost its dinner.

"Whether it was that we were just sitting down to our own picnic or not I cannot say, but the three geologists . . . only sympathized with the bird."

It was in 1836, after several years of lobbying from the scientific community (Professor Eaton in particular), that New York lawmakers approved $23,000 dollars each year for the extensive, 4-year surveying effort it would take to create the "Geological and Natural History of New York." It was intended to be a comprehensive assessment of the state's natural resources in geology, botany, zoology, paleontology, and mineralogy. At Eaton's suggestion, then-governor William Marcy (1786–1857) divided the state into four districts. John Gebhard Jr. was appointed as one of the six assistants to serve the first district, which included his hometown, Albany's Capital District, and portions of the Hudson River Valley.

The first volume of *Geology of New York* in 1843 is filled with numerous contributions by the Gebhards. They include geological reports, numerous fossil descriptions, and illustrated cross-sections of the soil and rock layers (stratigraphy) at several Schoharie County locations. A map by John Jr. of Ball's Cave is included. An illustration of the entrance to Howe's Cave is also included in this volume, although the cave's discovery at that time was so recent that a detailed description was not available by the publication's deadline.

The Gebhards' work for the massive geological survey found its way into other publications as well, notably Professor Hall's books on paleontology—the definitive, beautifully hand-illustrated nineteenth-century work in that field. The eight-volume *Paleontology of New York* (1847–1894) includes John Jr.'s *Trochorceras Gebhardi*, an ancient marine mollusk that inhabited the tropical sea that was Schoharie County 265 million years ago.

The huge collection of fossils, rocks and minerals, botanical samples, and other state resources that flooded into Albany from across the state during the survey formed the basis of today's New York State Museum. In 1845, a curator was hired to catalog the specimens being held in cramped quarters at what was called "Old State Hall."

In 1849, John Jr. was appointed curator of this natural history "cabinet," which was the quaint name for the collection. He was the second to hold the post. The State Regents noted in its annual report for 1850, "Great reliance is placed on his well-known devotion to, and his knowledge of, Natural History; and he has already given an abundant earnest in the industry and zeal with which he has entered on the engagements of his office." Remember, Junior was self-taught. He served through 1856, when he was 54.

It was also during this period that John Jr., in 1853, sold his cave to William H. Knopfel, who again renamed the cave—to Knopfel's Cave. (Knopfel, recognizing there was money to be made in the cave tourism trade, planned a hotel above the cave with a circular staircase descending the entrance shaft. His plans were never realized, and the cave remained in its natural state.)

In 1853 and again in 1873, the state purchased portions of the extensive Gebhard collection of fossils, cave formations (!), minerals, and other natural curiosities, more than 28,000 items overall. The purchase price together was about $5,500. Portions of the Gebhard collection have been sent to museums worldwide, but the New York State Museum continues to hold most of the Gebhards' lifetime collection.

Sadly, among what the museums gained from the caverns of Ball's/Gebhard's/Knopfel's (now Gage's Cave), Schoharie County and the caving community lost. Times were different then.

"Tons of rare minerals have been removed from the several rooms of this cavern," Historian Simms documented as early as 1845,

> . . . to adorn the cabinets of practical geologists. Stalactites and stalagmites, of semi-transparent alabaster, white as Alpine snow, and of every seeming variety of shape, have been taken from this laboratory. Minerals depending from the ceilings, or attached to the walls and floors, were removed by the early visitors, but many of the richest specimens have been discovered at a later period, by digging in the earthy floors. Some of the slabs of alabaster . . . are found to contain geodes filled with beautiful thread-like crystals.
>
> A specimen weighing several hundred pounds now adorns the valuable cabinet of John Gebhard, Esq., which was removed

by immense labor from the music saloon and drawn to the surface by a windlass. It is a mass of pure white alabaster, which has incorporated in its formation several stalagmites, and projecting from a part of which are forty-one distinct stalactites of various sizes, pointing, like so many magnets, to the center of all gravity.

Most of the Gebhard collection has likely been stored away in boxes and straw-filled crates to make room for later museum acquisitions. One item in particular has titillated the imagination of geologists, cavers, historians, and teenage boys since first described by Simms, who likely suffered pareidolia[1]:

It is a female bust, or rather breast, of purest alabaster; the contour is French, and approximates surprisingly to nature, on which account it is one of the most valuable of all stalagmitic formations—for it is a form which may be admired without the fear of its imbibing false pride or blushing at the exposure of its own charming proportions.

John Sr. died in 1854; Junior died in 1889. Both are buried in the Schoharie Lutheran Cemetery, just off Main Street.

"A single candle reflects but a sickly light in this dungeon of nature, but the writer once visited it when some thirty other individuals were there on the same errand, and the light of thirty torches discovered the magnificence of the apartment."
— Historian Jeptha Simms, in 1845, describing a visit to Gebhard's Cave

---

1. Pareidolia is the psychological term for seeing something in something else, e.g., Lincoln's face on a piece of toast.

1889 Photos Offer Rare Look Underground at Old Howe's Cave

Images by Noted Adirondacks Photographer Capture Lost Underground

Howe Caverns has always attracted its share of notable guests. As one of the earliest caves in the United States to open to the public (in 1842), many of the first visitors were the scientists, naturalists, and travel writers of the period.

Curious explorers came from around the world. Some wrote quasi-scientific papers or documented significant geological findings, but most described more fulsome, adventurous accounts of their eight to ten hours exploring old Howe's Cave. Several historical accounts survive.

Perhaps one of the most valuable historical records of Howe's Cave before its modern development comes not from a writer's pen, but from the photographer's lens—that of Seneca Ray Stoddard (1843–1917). He is most often credited and most familiar as S.R. Stoddard.

The Chapman Museum in Glens Falls houses more than 3,000 of Stoddard's photos. The museum writes:

> Stoddard is best known for his photographs of the Adirondack Mountains, but he also was a cartographer, writer, poet, artist, traveler, and lecturer. A sign painter by training, he turned to photography in his twenties. From his business base in Glens Falls he

carried his cameras throughout the region, capturing the vistas and scenes of Adirondack life over a span of forty years.[1]

At age forty-six, Stoddard lugged his photo equipment—including a heavy box camera, wooden tripod, cowl, dozens of heavy glass photographic plates, explosive flash powders, and accessories—to Howe's Cave in what must have been the early months of 1889. (His photos show that there is no snow on the grounds of the Pavilion Hotel, but inside the cave entrance are several ice formations that could easily be mistaken for stalactites and stalagmites.)

An assistant, who operated Stoddard's portable darkroom, was likely with him, according to Chapman Museum Director Timothy Weidner. He also hired a local—in this case probably a tour guide—to act as a "runner" to carry the imaged glass plates from photographer to the darkroom, a light-proof tent containing the trays of chemicals used to reveal the captured images.

At the time, Howe's Cave was operated by Joseph H. Ramsey's Howe's Cave Association, which had purchased the cave from the Howe family about 20 years previously. The association had made some improvements,

Noted Adirondacks' photographer and naturalist Seneca Ray Stoddard of Glens Falls circa 1880, courtesy of the Chapman Museum, Glens Falls. Stoddard published a 40-page booklet of his photos of old Howe's Cave in 1889. Several are used in this book.

such as clearing paths and installing gas lighting in some sections of the cave, but it was still no easy trip—especially carrying an estimated 30 to 50 pounds of photo gear.

Photography was not new; the first permanent images captured by camera were from France in the 1830s. However, flash photography—using artificial light to capture an image—was relatively new. And, underground in pitch black, it was *all* flash photography.

Until the 1900s, photographers like Stoddard used a small mixture of explosive magnesium and potassium chlorate powder. Held on a "flash stick" platform, a cable connected it directly to the camera, usually with the photographer holding it aloft so that when the camera's shutter actuated, it triggered the bright, flash-powder explosion.[2] By connecting several flash sticks together by cable, Stoddard would have been able to light some of the cave's larger rooms.

Inattentive photographers often suffered serious burns using this method. In the close quarters of the cave, the noxious smoke from the flash must have lingered for many hours.

Judging from the photos, Stoddard took the full tour. Without stopping to compose and painstakingly labor over a photograph, the tour in that era took an estimated five hours from the entrance to the Great Rotunda at the end of the cave, and then five hours back. Stoddard captured 22 images in the cave overall, and another photo underground at Ramsey's cement mine, which was nearly adjacent. It likely took several exceptionally long days to complete the on-site photography, and many more days to carefully expose the glass plates in the on-site darkroom. Stoddard's wife, Helen, made prints from the images on an upper floor of the couple's Glens Falls home, according to Museum Director Weidner.

Fortunately for historians, many of Stoddard's photos are of sections of the cave that no longer exist, as portions nearest the cave entrance were destroyed when the cement industry expanded. Stoddard photos include such colorfully named cave features as "Martha Washington's Hood," "Washington Hall," "Tall Man's Torment," "Music Hall," "The King's Corridor," "The Devil's Gateway," and "Pulpit Rock," to name a few.

Sections of the cave that remain part of today's cave tour are easily recognized, despite their modern accoutrements.

"Stoddard had a wonderful eye," said Dick Danielsen of Dick Danielsen Photography, in Warnerville, Schoharie County. "His photographs have excellent composition. Proper lighting is exceptionally difficult underground, where the cave walls seem to absorb all light.

"His photos capture the beauty of the cave and are visually informative. That all this was done with very heavy and bulky equipment is amazing," said Danielsen, whose own 30-plus-year career in Schoharie County has included images taken both above and below ground.

The "Giant's Gallery," by Seneca Ray Stoddard. This 1889 scene is today called the Pool of Shiloam, or Pool of Peace.

1889 Photos Offer Rare Look Underground at Old Howe's Cave

Available Online

Stoddard's final product was printed later that year as *Howe's Cave*, by Nims & Knight Publishers, located in the now-historic Cannon Building at Monument Square in downtown Troy, New York. The 40-page booklet is 6 ⅜" by 8 ⅝"; a PDF version is available online through the Library of Congress: archive.org/details/howescave00stod.

1. "Stoddard was a great advocate of conservation of the Adirondacks. One of his drawings, portraying drowned lands caused by damming streams, was to document what he considered to be the uncontrolled impact of man. In 1892, his persistent lobbying of the New York State Legislature was rewarded when Governor Hill signed a bill establishing the Adirondack Park." The museum maintains an online database of Stoddard images.

2. "Quick lesson in history—flash photography," from the Moneymaker Photography Web site.

3

Here She Caves ... Miss America

And One Man's Attempt to Open a Schoharie Cave

July 26, 1958—Wearing elegant white gloves, high heels and carrying a bouquet of roses, Miss America Marilyn Van Derbur, accompanied by Miss New York Miriam Sanderson and Miss Schoharie County Marion Gage, helped dedicate Schoharie County's third cave planned to open to tourists.

Schoharie Caverns, owned by Attorney James L. Gage, of Esperance, must have been closed to the public the following day. Gage, however, continued for many years to expand the cave's potential appeal by extending its size. He also constructed a cabin to welcome visitors and landscaped the premises for parking and camping.

The much more famous Schoharie County caves—Howe Caverns and Secret Caverns—opened in 1929 and have been money-makers ever since. Schoharie Caverns, between Schoharie village and Gallupville, was the next likely contender for commercial success. In fact, the Mallery family, which owns Secret Caverns, tried to make a go of it in 1935.

It seems unlikely that misses Van Derbur, Sanderson, and Gage (Jim's daughter) took the cavern tour that day, or at least not until after their photo was taken. Caves, even with paths cleared, electric lights, and boardwalks, aren't great for high heels and white gloves. Readers can imagine the effects the tour's high point, the "Shower Dome," might have had on immaculately coifed hair and white gloves.

Here She Caves . . . Miss America

Miss New York State, now 82-year-old Miriam Russell of Troy, doesn't recall that particular sunny day over 62 years ago. Like other state pageant winners, she was paired often with Miss America for a variety of special events. Twenty-one at the time, "that year was just a whirlwind," she said. After the dedication ceremonies at the cave, the young ladies were off to a thrilling stock car race at Fonda Speedway. Both events, sponsored by Pepsi-Cola, were promoted by Gage, who was also a co-owner of the Fonda racetrack and planned to sell tickets for the races at the cave.

This nearly forgotten, almost unknown, story of glamour, pageantry, and cave exploring came to light only a few years ago. Gage, who served

Front left, Miss America Marilyn Van Derbur of Denver on the boardwalk during the July 26, 1958, grand opening ceremony for Schoharie Caverns in Shutters Corners. She is followed by Miss New York Miriam Sanderson of Rensselaer, a Mr. and Mrs. Fitzgerald, Miss Schoharie County Marion Gage of Esperance, and other, unidentified, dignitaries. The photo is from a color negative in the James L. Gage files, possibly taken by Russell H. Gurnee, a longtime friend of the Gages.

as Schoharie County's district attorney (1941–47), had a lifelong interest in caves. He was an explorer, early member of the National Speleological Society, and friends with society president (1961–63) Russ Gurnee and his wife Jeanne, both of New Jersey.

Following Gage's death in 1991, Jeanne Gurnee received several large cardboard boxes of Jim's letters, publications, photos, and other material on Schoharie County caves from his widow, Sally.

In October 1997, Jeanne passed the material on to former Cobleskill resident Bob Addis and Emily Davis of Speleobooks in Schoharie. Jeanne wrote: "They are a fascinating history and show the thoroughness of Jim's attention to the details of the caves of the Schoharie area. I know you will take good care of them and pass them on to another good custodian when you think the time is right."

The boxes sat in Davis's unheated barn on Barton Hill until January 2013, when Addis took them to the Troy home of Chuck Porter, another long-time caver. "We were surprised by the spelean treasures we uncovered," said Porter. The pair sorted through and organized the material. Porter has published several articles based on the collection that have appeared in regional and national cave-related publications, notably the quarterly *Northeastern Caver*, which he edited for decades.

Jim Gage had been interested in Schoharie Caverns for a number of years. The cave is generally well-known in the Gallupville area and has had several different names; most are after the landowner who farmed the fields around the cave, which provides a steady source of water. It has been called Spadeholt's, Nashbolt's, Cook's, Treadlemire's, and Shutters Corners' cave, according to caver/author Clay Perry in his 1948 book, *Underground Empire*.

Gage apparently visited the cave in September 1957. After Russ Gurnee visited the cave, he wrote to Gage in January 1958, "We saw written in the mud on the bottom of a sunken boat, 'J.G., September 1957,' and I think I recall your saying you had gone partly into the cave."

Gurnee (1922–1995) wrote to Gage that it was a "very fine and unexpectedly large cave. You should really visit this cave, as it is quite easy to enter and with a minimum of wading you can get to the dry portions in a very short time. [It] is a splendid example of a New York State meandering stream passage cave."

After writing that the cave was "a horrible example of misdirected effort in commercialization," Gurnee cautioned, "The cave layout is not conducive to commercial development."

Gage apparently thought otherwise. He announced the purchase and renamed it "Schoharie Caverns." In a May 7, 1958, letter to his caving associates, he described his plans to make improvements: "We are expecting that the entrance will be opened by backhoe during the present week so that one may walk erect for the first 660 feet . . . we are also expecting to have a shelter constructed which would be suitable for overnight camping and to have electricity available in the small camp."

While the cave may never have been fully open to the public, it was always open to members of the National Speleological Society, who were invited to explore beyond the developed portion of the cave, at their own risk.

Over the next five years, Gage continued his labor of love to present the cave to the public, both above and below ground. In the cave, efforts failed to lower the water level at the cave's end; doing so would have exposed additional passage that would have doubled the length of the 2,000-foot-long cave.

But enthusiasm for the project waned. Gage placed an ad in April 1961 in a New York City newspaper to sell the cave. He noted the amenities and all that had been accomplished: "15 acres of lands with cave rights on an additional 145 acres together with sign rights; electric wiring for 500 feet in the cave, a frame building with porch overlooking the valley, and a 16-foot driveway extending to a 75-car parking lot, both graded [with shale]."

Gage spent more than $11,000 to prepare Schoharie Caverns for its 1958 opening, and between 1958 and 1963 he spent nearly $14,000, according to his personal records. (That's equal to about a $250,000 today.) He wrote to one prospective buyer that more than 50 percent of the work to open the cave had already been completed.

Gage received seventeen responses to his ad, but never found a buyer. He advertised the "ten-year lease was to be $5,000 for the first year plus 10 cents on each adult admission, thereafter $1,000 each year plus 10 cents on each adult admission after the first $10,000 of adult admission charges. [The] minimum capital requirement [was] $10,000."

As Gage's plans for a tourist attraction in Shutters Corners faded, the property around the cave began to revert to its natural state. But Schoharie Caverns remained a Mecca for northeastern cavers, who explore, camp, and hold conventions there on occasion. Volunteers maintain the site. When the picturesque little cabin overlooking the valley was destroyed by fire in 1993, cavers built a new one.

After Gage's death on July 11, 1991, at the age of 83, Mary and Jennifer Gage donated the caverns property to the National Speleological Society in 1994 to become the Schoharie Caverns Nature Preserve. A bronze plaque was mounted over the cave entrance in 1997 commemorating Jim Gage and Russ Gurnee's "considerable contributions to northeastern caving." (The plaque hangs near the iconic entrance gate to Schoharie Caverns, built by Jeanne Gurnee. The wrought-iron gate—a spider on its web chasing a fly—occasionally leads to some confusion with "Spider Cave," another, smaller cave in the Gallupville area.)

While the celebration and glamour of the July 26, 1958, visit by Miss America may not have created long-lasting memories among cave visitors, it did among some of those involved. Following up with a request by a Columbia University scientist for a tiny cave specimen to date radioactively, Gurnee wrote to Gage, "Dr. Broecker may prefer to date stalagmites; I would prefer to date Miss America."

4

From "Nameless" to "Secret"

Flashlight Company Shines a Light on the "Lesser Caves" of Schoharie County

A Russian immigrant who sold his handmade batteries by pushcart on the streets of New York City in the early 1900s had the unlikely role of helping bring a famous Schoharie County attraction to light.

In the second half of 1928, there was considerable interest growing throughout the Northern Catskills, Capital District, and beyond for the planned reopening of world-famous Howe Caverns. The cave discovered in 1842 by Lester Howe and once host to visitors and scientists from around the world had been closed for several decades. A corporation formed in 1927 was hard at work building an entrance lodge and installing an elevator, electric lights, and stone pathways.

In the midst of this renewed interest in caves, Arthur Van Voris, whose family owned a Cobleskill hardware store, assembled a team of amateur explorers for a newspaper series on other caves in Schoharie County. Accompanying Van Voris were Cobleskill Post Office employee Edward Rew and the "VanNatten boys," who owned farm property in the cave country. Over the course of several weeks, the Van Voris team of explorers filed reports on four local caves: Ball's Cave, outside of Schoharie; Benson's Cave and Nameless Caverns, both near Sagendorf Corners; and Selleck's Cave, Carlisle.

It is not known what brought together this adventurous group of explorers and the Hoboken, New Jersey-based Bright Star Flashlight Company,

Partnering with a team of Cobleskill-area explorers, the Hoboken, N.J.-based Bright Star Battery Company used the excitement around the 1929 reopening of Howe Caverns to help sell their products. The resulting brochure describes the first exploration of what is today Secret Caverns.

but what better way to promote flashlights than having them illuminate the total darkness deep inside the earth? Van Voris & Sons Hardware, on Main Street, Cobleskill, was an authorized distributor of the brand. It seems likely that Van Voris oversaw equipment needs for his fellow explorers, and it's reasonable to assume he picked up the Bright Star brand from off the store's shelves.

"Like a Star from Heaven"

Russian immigrant Isadore (Israel) Koretzky arrived at Ellis Island early in the 1900s, according to a company history. He found work with a Manhattan firm that manufactured dry cell batteries.

Working after hours in the basement of his tenement building on Brite Street, the ambitious Koretzky developed improvements to the basic dry cell battery and, in 1909, he founded the Bright Star Battery Company, combining his street name with a boast that "the batteries produced a light like a star from heaven." The company grew quickly.

Surely, Koretzky thought the public interest in the caves of upstate New York would pay off for Bright Star. Using Van Voris's newspaper series as starting point, a team of Bright Star ad agency copyeditors, illustrators, and photographers created "Cave Exploring with Bright Star Flashlights," a sixteen-page, pocket-sized brochure that invited readers to "enjoy the interesting story of these intrepid explorers of the almost unknown caverns in Schoharie County, New York." Bright Star invested in a quarter-million printed copies of the two-color brochure and sent them to distributors across the country.

"Cave Exploring with Bright Star Flashlights"—now extremely hard to find—describes the local group's explorations and is attributed to Van Voris. The brochure was intended to sell flashlights and batteries, and the text dutifully relates how Bright Star products performed with "eminent perfection—always disclosing some new and entrancing detail at each turn of the light."

Nameless No More

The brochure describes Van Voris's excitement with the Nameless Cavern, the entrance of which lies about a mile to the northeast of the Howe Caverns' entrance lodge. "Words are all too inadequate," he said, "to describe what we saw in Nameless Caverns.

"Shooting the brilliant rays of the Bright Star out through the darkness, a solid rock wall looms up ahead and one would conclude this to be the end of possible exploration—but at the very bottom of the wall is a small opening just high enough to crawl into on hands and knees . . ."

The brochure, though not printed to be a historical document, does offer bits of local history. Unlike Howe Caverns, the discovery of Nameless Caverns was not heralded to the world. The large sinkhole entrance was probably known by hunters and farmers in the area for many years.

An adventurous member of the Van Voris team squeezed through the tight passage beyond the rock wall that blocked their progress. Van Voris reports, and the Bright Star brochure documents, "It is our belief that one of our group, our guide on this trip, was the first to fully explore the cavern." Sadly, the name of that ardent explorer is lost to history.

Nameless Caverns finally got a name not long after Howe Caverns opened to the public with great fanfare on Memorial Day, 1929. Roger Mallery, the civil engineer responsible for much of the construction work at Howe Caverns, began work in the winter of 1928 to open the nameless cave Van Voris had been so enthralled with, thanks in part to the illumination provided by Bright Star flashlights and batteries.

Now a third-generation family business, Secret Caverns opened in August 1929. Its first brochure describes the origin of the new name: "Years ago, Lester Howe, the discoverer of Howes Cave, used to remark that he knew of a wonderful cavern nearby but insisted on keeping its location a secret." It is open daily April–December.

Secret Caverns tour guides sometimes mention that a "flashlight company's publicity stunt" was also instrumental in the cave's development.

It is interesting to speculate on the role Van Voris and the Bright Star company may have had on the decision to develop Secret Caverns to the public. Was it the added publicity? Was Mallery an unnamed member of the Van Voris party?

The Koretzky family sold the Bright Star company in 1955. It changed hands a few times until February 1998, when it was acquired by Koehler Manufacturing of Marlboro, Massachusetts. The company was renamed Koehler-Bright Star, Inc., and operations were moved to Wilkes-Barre, Pennsylvania. It still manufactures flashlights and batteries today.

Arthur Van Voris published his newspaper series, without the Bright Star embellishments, as "The Lesser Caves of Schoharie County." He maintained a lifelong interest in caves and in local history, served as editor of the Schoharie County *Historical Review*, and wrote a weekly column on history for the *Times-Journal* of Cobleskill.

5

Blenheim Monster Serpent Mystery Unsolved

Beast Slithered from Its Mountain Cave

Two young men in the Town of Blenheim in southern Schoharie County, ages twenty and fifteen, were confronted by a "huge monster" that hovered six to eight feet above their heads as they rested on a rock ledge following a fishing trip.

The young men were seated on the north side of a steep mountain in which, sources wrote, there are several openings or caves, "the interior of which never has, and since (the monster's sighting) . . . probably never will be explored by any human being."

The names of the young men were not fully released and, although badly shaken, they were not injured. After recovering from their initial fright—and several loud shrieks—they clutched one another high on a rock ledge where they watched in terror as the monster, undaunted, continued his course down the hill for some distance and then turned back into his den.

"After making all due allowances for their fright," reports noted that the young men "are fully convinced that [the creature's] body exceeded thirty feet and his body was as large as a common saw log." They also said it "was covered with irregular spots—about the size of a man's hand—of bright red and jet black."

To residents of the area, there's no need to panic. The incident described took place in mid-July 1826 and was reported by *The Schoharie*

Republican, which published Wednesdays from 1819 until 1854. This rare report was picked up and reprinted on August 15, 1826, in the *American Mercury* of Hartford, Connecticut, where it was found unexpectedly by the author. The *Mercury* was published 1784–1833.

The day after the incident, the report continued, the oldest of the pair, "said to be no ways deficient in courage, went alone to the den and seated himself on a rock that projected over the [cave] entrance." It was not long when "his Snakeship" made his appearance. The creature descended farther down the mountainside than it had before, "breaking the old limbs and sticks as he passed over them."

The young man, whose name the paper "believed to be Sandford, made some noise from above, then he [the giant serpent] turned his course and again entered the den."

The newspaper reported, "The alarm was given in the neighborhood, and a number of people collected at the den, but they were not gratified with a view of the monster; his course, however, could be plainly traced, and where he had passed over the sand or leaves, it appeared like the trace of a large log.

"Since these facts have spread, the cave has been visited by a great number of people from the adjoining town of Jefferson, many of whom have visited the cave and seen the track of the serpent." (We wonder if the John Gebhards [see previous chapter] were among them.)

Apparently, no one in either Blenheim or Jefferson thought the sighting was a hoax. The paper went on to report that neighbors had "the utmost confidence in the veracity of the young men who saw him, and as an additional confirmation, they say that a very offensive smell similar to that of the large snake has been observed by all who have visited the place."

There is no reference to the Blenheim monster serpent in either of the two nineteenth-century histories, Jeptha R. Simms's 1845 *History of Schoharie County* or William C. Roscoe's *History of Schoharie County 1713–1882*.

The 200-year-old tale also stumped Schoharie County Historian Ted Shuart and Town of Blenheim Historian Liz Arrandale. Further, any amateur geologist can verify that there are no caves in the hills around Blenheim. No one remembers finding the remains of a giant serpent during the 1969–73 construction of the Blenheim-Gilboa Power Project. If they did, they didn't tell the news media.

Newspaper hoaxes were used occasionally in the mid-nineteenth century to build readership, but not as early as 1826. *The New York Sun* newspaper's "Great Moon Hoax" was among the first, in 1835, and the better-known "Cardiff Giant hoax" wasn't until 1857.

It seems like the mystery of the Blenheim Monster Serpent, like Big Foot or the Loch Ness Monster, will remain unsolved for the time being. But it makes a good yarn in the hills of Schoharie County.

Published Sources

Sources used in Sections I, III and IV of this book were directly attributed in the text for the convenience of the reader. Much of the text in Section II was taken from old company pamphlets, newsletters, and publications held by private individuals. We include them here.

Biographical Review Volume XXXIII: Containing Life Sketches of Leading Citizens of Greene, Schoharie and Schenectady Counties, New York (Boston: Biographical Review Publishing, 1899).

Cobleskill *Times-Journal*. Various issues 1931–47. Held at the Cobleskill Public Library.

Cudmore, Dana. *The Remarkable Howe Caverns Story*. 2nd edition. (Woodstock, N.Y: The Overlook Press, 2001).

Danforth, Pierre. *Schoharie County Directory*. (Middleburgh, N.Y.: 1899).

Fake, Kenneth. *History of the Town of Cobleskill*. (1937).

Keyes, Donald. *Eli Rose, the First Superintendent of Howe's Cave Cement Company*. A publication from Stone Fort Days, Schoharie, 1986.

Noyes, Marion J. *A History of Schoharie County*. 1964.

Roscoe, William E. *History of Schoharie County*. (by 1882). Accessed at the Schoharie County NYGen Web site.

Safety Flashes, a publication of the North American Cement Company, issues in 1941–42, held by the Cave House Museum of Mining and Geology.

Shutt, Donald. *Schoharie Days*. Undated (approx 1961) publication of the Marquette Cement Corp.

"Smallest Post Office?" *Schoharie County Historical Review*. Spring–Summer, 1977.

A Trip through the North American Cement Plant, Howes Cave, N.Y., a pamphlet published 1931 by the North American Cement Company. From a private collection.

Index to the Cast of Characters in *Underground Empires*
The Explorers, Entrepreneurs, Promoters, Scoundrels, Scientists, Innovators, and Aficionados

Robert "Bob" Addis, former tour guide and caver, 23, 26, 80, 138, 218–220, 304

Fernando Boreali, early-twentieth-century quarry worker, 185, 186

Virgil Clymer, caverns' manager, 61, 68, 71–75, 101, 102, 156, 160, 162, 163

Floyd Collins, famous caver, 67, 68

Charles E. Dewey, Howe family member, 100, 145

Charles E. Dewey Jr., Howe family member, 270, 271

Hiram Shipman Dewey, Howe family member, 40, 64, 146

Amos Eaton, nineteenth-century geologist/professor, 292–294

James L. Gage, attorney/cave owner, 7, 120, 122, 302–306

Emil Galasso, president of Cobleskill Stone Products/a new owner of Howe Caverns, 229, 238, 259, 264

Sam Galasso, Howes Cave stone products' supervisor, 231, 234, 235

John Gebhard Jr., nineteenth-century naturalist/state museum curator, 9, 62, 175, 288

John Gebhard Sr., nineteenth-century naturalist, 9, 175, 288

Paul Griggs, geologist/education director at Cave House Museum, 232, 275, 279

Ben Guenther, founder/educational director of Cave House Museum, 221, 227, 242, 264, 265, 269, 279

Floyd Guernsey, quarry worker, 56

James Hall, nineteenth-century paleontologist, 292

Seth's Henry, murderous warrior, 50

Robert "Bob" Holt, former cave manager, 24, 250, 251, 258, 265, 271

Henry Homburger, caverns' surveyor, 78, 86

Rev. Horace Hovey, nineteenth-century geologist/author, 57, 147

Lester Howe, discoverer of Howe's Cave (Howe Caverns), 5, 16, 17, 19, 22, 41, 48, 49, 51, 53, 60, 61, 63, 64, 90, 93, 100, 104, 108, 127, 129, 132, 133, 135, 136, 139, 141, 142, 144, 145, 146, 153, 161, 163, 167–169, 176, 213–215, 226, 245, 252, 257, 267, 271, 276, 281–283, 292, 307, 310

Halsey John Howe, Lester's son, 10, 41, 64–66

Harriet Elgiva Howe, Lester's daughter, 10, 40, 42, 64, 66, 82, 145, 271

Helen Howe, distant relative, 144, 145

Herbert Howe, distant relative, 144

Huldah Ann Howe, Lester's eldest daughter, 10, 42, 65, 66

Lucinda Howe, Lester's wife, 10, 64, 176

Warren Howe, a distant relative/ genealogist, 22, 51, 64, 145

Morris Karker, early caverns' employee, 77, 81, 95

Isadore Koretzky, flashlight and battery maker, 308

Sir Charles Lyell, preeminent nineteenth-century geologist, 292–3

"R.J." Mallery, current manager of Secret Caverns, 118

Roger H. Mallery Jr., Secret Caverns owner/attorney, 78, 117, 165, 166, 213, 237

Roger H. Mallery Sr., developer of Secret Caverns (1929), 77, 81, 83, 84, 89, 94, 107, 108, 115, 116, 119–21, 135, 161, 163, 169, 285, 302, 310

T. N. McFail, early cave explorer, 9, 160

Clemens McGiver, created the "adaptive reuse" for the Howes Cave Quarry, 227–30, 253, 264–66, 268, 277, 279

Frances Howe Miller, distant relative of Lester Howe, 41, 65, 144

Miss America 1958, co-promoter of Schoharie Caverns, 302, 303, 306

Miss New York State 1958, co-promoter of Schoharie Caverns, 302, 303

John Mosner, proposed development of Howe Caverns, Inc., 67–69, 71, 72, 86, 90, 100, 160

Lavina Mulbury, former Howes Cave resident, 198

John Pangman, former Howes Cave resident, 191–193, 203, 204

Clay Perry, author/cave explorer, xxii, 60, 76, 109, 121, 130, 133, 134, 136, 139, 147, 149, 169, 213, 304

"Pip," early explorer, 26

Louise Provost, owner of Cave House/ boardinghouse, 82, 84

Charles Ramsey, Joseph's son, 46, 64, 183

Joseph H. Ramsey, railroad tycoon/ cement manufacturer, 43–47, 52, 53, 56, 62, 66, 168, 179, 183, 257, 265, 298

John Peter Resig, Rev., fictitious eighteenth-century cave resident, 7, 8

Edward A. Rew, early cave explorer, yarn spinner, 92, 109, 110 122, 129–132, 134, 141, 142, 169, 285, 307

Chauncey Rickard, created caverns' tour presentation, 94, 154

Victor Rickard, Schenectady pilot, 102, 103

Delevan Clarke "Dellie" Robinson, cave explorer/promoter/owner, 59, 71, 72, 79, 118, 122–124, 126–128, 135, 136, 169, 188, 282, 284

Eli Rose, superintendent of first Howes Cave cement quarry, 176–178, 183

Paul Rubin, caver/karst hydrologist, 214–217, 220–224, 242, 280

John D. Sagendorf; former general manager of corporation, 211, 212, 214, 215, 237, 239

Index

John J. Sagendorf, first corporation secretary, 69, 73, 80, 101, 103, 105, 240, 284

Mabel Sagendorf, John J.'s widow/long-time employee, 90, 105

Nancy Sagendorf, educator and long-time employee, 239, 240, 271, 285

Walter H. Sagendorf, first corporation officer, 67, 69, 72, 73, 100

Jonathan Schmul, fictitious eighteenth-century peddler, 7, 8

Anthony "Tony" Spenello, former quarry kiln supervisor, 190, 192, 201

Seneca Ray "S.R." Stoddard, noted Adirondacks photographer/naturalist, xvi, xvii, 15, 20, 25, 41, 96, 97, 180, 257. 297–301

Jim VanNatten, early caverns' guide, 95, 102, 138

Arthur H. Van Voris, hardware store owner/caver/author, 91–93, 109, 110, 117, 122, 129, 131, 132, 134, 163, 169, 174, 175, 284, 301–310

Owen Wallace, caverns' electrician, 101–103, 160

Henry Wetsel, Lester Howe's neighbor and cave property owner, 12–15, 17

Charles Wright, a new owner of Howe Caverns, 238, 239

T.L. Wright, former Penn-Dixie employee, 196

E.F. Yates, journalist, 9, 16, 39, 175

Two '50s-era promotions for Howe Caverns. Pennants were popular, and the "Interstate Road Guide" brochure unfolds for a detailed road map of New York and surrounding northeastern states. It features full-color artwork of many of the cave's features.

Black Dome Press

Black Dome Press publishes New York State and New England histories and guidebooks, with particular focus on the Adirondacks, Hudson River Valley, Catskill Mountains, Capital & Saratoga Regions, the Shawangunks, Mohawk Valley, the Berkshires & Taconics—the scenic and storied lands of Rip Van Winkle, John Burroughs, Thomas Cole & the Hudson River School of landscape painting, and the domains of influential American families like the Roosevelts, Livingstons & Van Rensselaers. Topics include: nature & hiking, art history, Dutch and English exploration & colonial life, geology, preservation and conservation, Native Americans, homesteading, railroads, the grand hotels and great estates, folklore and folk art, genealogy, the Civil War, American Revolution & French and Indian War, community and city histories, waterfalls, kayaking, architecture, historic sites and museums, and nature photography.

Black Dome Press is located in the seventeenth-century Dutch village of Catskill, New York, where the Hudson River meets the Catskill Mountains. For more information, visit **www.blackdomepress.com** or call (518) 577-5238 for a free catalog.

A clean-cut Howe Caverns tour guide welcomes visitors in a 1950s-era brochure.